W9-CLZ-493

LEADERSHIP AND URBAN REGENERATION

Volume 37, URBAN AFFAIRS ANNUAL REVIEWS

LEADERSHIP AND URBAN REGENERATION

Cities in North America and Europe

Edited by

Dennis Judd
Michael Parkinson

Volume 37, URBAN AFFAIRS ANNUAL REVIEWS

For information address:

SAGE Publications, Inc.
2455 Teller Road
Newbury Park, California 91320

SAGE Publications Ltd.
6 Bonhill Street
London EC2A 4PU
United Kingdom

SAGE Publications India Pvt. Ltd.
M-32 Market
Greater Kailash I
New Delhi 110 048 India

Printed in the United States of America

Library of Congress Cataloging-in-Publication Data

Leadership and urban regeneration: cities in North America and Europe
/ [edited by] Dennis Judd, Michael Parkinson.
p. cm.—(Urban affairs annual reviews; v. 37)
Includes bibliographical references.
ISBN 0-8039-3980-9.—ISBN 0-8039-3981-7 (pbk.)
1. Urban renewal—United States—Congresses. 2. Urban renewal—Canada—Congresses. 3. Urban renewal—Europe—Congresses. 4. Community leadership—United States—Congresses. 5. Community leadership—Canada—Congresses. 6. Community leadership—Europe—Congresses. I. Judd, Dennis R. II. Parkinson, Michael. III. Series.
HT108.U7 vol. 37
[HT178.U6]
307.76 s—dc20
[307.3'416'097] 90-43946
CIP

FIRST PRINTING, 1990

Sage Production Editor: Susan McElroy

Contents

Foreword

IN THIS CENTURY, both cities and leadership have had mixed reputations. For some, leadership does not sit comfortably with democracy. We have seen too many Führers, beloved Chairmen, and parties with leading roles. For many, cities have been the breeding grounds of crime, poverty, and impiety. Leaders like W. H. Lever, in whose picturesque factory restaurant this conference was held, led their troops firmly away from the urban centers, into model villages, new towns, and the mass transit suburbia.

Strangely, in Anglo-Saxon usage, it is more natural to contrast leadership with autocracy and demagogy than to assimilate it. In this more optimistic paradigm, it is by leadership that individuals bring larger groups to the achievement of common goals, without coercion, bribery or deception, subordinating personal interests to group needs. In these clothes, leadership is seen as an essential element of democracy.

The city has had its enemies from the earliest times. But there has been another view. Cities have been great centers of innovation and learning, where science and administration, commerce and the arts, have lived and worked together. Citizens have built fine buildings, rich gardens, wide avenues, and quiet squares. Our cities have been gregarious, exciting places, where cultures and philosophies compete, where icons are broken, worshipped, and conserved.

When times are good and things change slowly, the need for leadership may not be great. In times of adversity, crisis, or rapid change, society moves from where it was to some other place, by choice or by force of events.

Throughout the developed world, the great nineteenth-century cities have been in crisis. Evils, which in times of growth existed alongside wealth and progress, multiplied, surrounded by dereliction.

It may be that the market decides—one way or the other—the future of our cities. But the tradition of the city is to take a hand in its own destiny. To steer away from adversity means to identify options, declaring a vision of what the future might be, defining the steps from the present to the vision, and leading people through those steps.

Many cities have found leaders, and many have turned adversity into success. We do not adequately understand either the circumstances that lead to success, or those that produce or inhibit leadership. We do not know what institutional frameworks allow leaders to be effective in cities. We do not understand sufficiently how the interests of constituencies, like investors, shoppers, or the possessors of vital skills, can be built into representative systems or how administrative structures can work with informal networks.

To make leadership in the cities more effective, we must know more about cities and more about leadership. The essays in this book make a unique and important contribution to the knowledge, and to the future, of one of man's greatest inventions.

W. G. Byrnes
Port Sunlight Village
Merseyside

Preface

THIS BOOK IS THE PRODUCT of a conference sponsored by the Centre for Urban Studies of the University of Liverpool in November 1989 at Hulme Hall, Port Sunlight, Merseyside. The location was entirely appropriate since it took place in the model village on the original estate of Lord Leverhulme, and the conference themes directly reflected the research interests of the Centre, which is supported by the Leverhulme Trust. The Centre is dedicated to a comparative exploration of the processes of urban regeneration in the belief that an informed public debate can make a contribution to the regeneration of the city in which it is based—Liverpool. It organizes a wide variety of academic and policy-oriented activities, which join the public, private, and community sectors in Liverpool with their peers from cities in Europe, America, and beyond to discuss their experiences and learn from their respective successes and failures.

This conference brought together academics from five countries to discuss a major theme of the Leverhulme Trust research agenda—the contribution of local leadership to urban regeneration. The cities discussed in this book all reacted differently to their problems and opportunities in the 1970s and 1980s. Academics who lived and worked in each city and knew them well were invited to analyze responses to decline, and the role of leadership in defining such responses.

Identifying the appropriate contributors to such a volume would have been a daunting task were it not for the fact that we are very lucky to be a part of a very lively community of scholars. Many previous conferences and events have provided occasions for the intellectual exchanges that prepare the ground for a collaborative effort such as this one. Michael Parkinson has been in the United States on several occasions, delivering papers and lectures and consulting with his colleagues. One of his visits to St. Louis was made possible by the Chancellor's Visiting International Scholar Program of the University of Missouri-St. Louis. Dennis Judd is an adviser to the Leverhulme Trust grant and has been supported by the Trust on his visits to Liverpool. The individual contributors to this volume form part of a growing community of transatlantic urban scholars who have been brought to Liverpool to share their experiences and perspectives with the Centre and its constituencies.

The Port Sunlight Conference was an extraordinarily rewarding and enjoyable gathering. For three days the participants shared their views and their research, and on occasion we debated sharply. Leadership is an elusive phenomenon. We found it relatively easy to discuss the restructuring of the world economy and the changes in national policies toward cities. We were able to describe the impact of economic change on cities, since cities on both sides of the Atlantic have been affected in remarkably similar ways by global economic restructuring. But the cities discussed at the conference responded to the changes in very different ways. What role did leadership exert in defining the responses? Answering that question turned out to be difficult indeed.

Perhaps this is one reason why the conference was so unusually rewarding. The role of leadership has been neglected for some time by scholars in our field, in part because the task of describing and documenting the immense changes that have been taking place in cities has been an important and daunting undertaking on its own terms. But such documentation is potentially misleading because cities are not simply passive objects, subject to external economic and political influences. To a considerable degree they can control their own fate. As social scientists we are interested in assessing such questions as "how much" and in "what direction" cities can be lead. The people who actually take on the role of urban leaders necessarily are interested in such questions as well, and therefore the answers are immediately compelling. The studies in this book are addressed simultaneously to a community of scholars and to a community of urban leaders.

Many organizations and individuals contributed to the success of the conference. In the largest sense the Trustees of the Leverhulme Trust and its Director, Sir Rex Richards, made the venture possible by so generously supporting the research of the Centre. The Nuffield Foundation and the Research Development Fund of the University of Liverpool provided invaluable financial support. The Vice-Chancellor of the University of Liverpool, Professor Graeme Davies, made a major contribution with his unfailing personal support for the work of the Centre. The entire administrative burden of the conference was superbly handled by Jean Parry in Liverpool. In St. Louis, Jan Frantzen efficiently and patiently helped bring the manuscript together, often enough typing on the manuscript after office hours. Two work study students, T. J. Sanders and Gary Garufi, burned up the fax lines between Liverpool and St. Louis.

A most indispensable contribution to the entire enterprise was made by the Chairman of the Centre's management committee, Bill Byrnes, Managing

Director of UML, part of Unilever plc. He encouraged the approach to the Leverhulme Trust that resulted in the founding of the Centre. He constantly challenged its Director to be ambitious and make no small plans. His vigorous commitment to intellectual inquiry and to the public life of the city of Liverpool is living testimony to the important of urban leadership. This book is dedicated to him.

Michael Parkinson
Centre for Urban Studies

Dennis Judd
University of Missouri-St. Louis

March 1990

1

Urban Leadership and Regeneration

DENNIS JUDD
MICHAEL PARKINSON

ECONOMIC CONSTRAINT AND POLITICAL CHOICE

Since the 1970s the scale of economic decline in cities in both Europe and North America has provoked a debate about the future of cities, and whether they could, or should, be "saved." On the one side were those who argued that it was unwise or impossible to resist the logic of the market forces that had plunged older cities into decline. For instance, in its 1980 report, the President's Commission for a National Agenda for the Eighties recommended against national policies designed to protect communities from economic decline. Indeed, the commission considered uneven growth among cities and regions as a sign that the U.S. national economy was successfully adapting to change. "It may be in the best interest of the nation," it suggested, "to commit itself to the promotion of locationally neutral economic and social policies rather than spatially sensitive urban policies that either explicitly or inadvertently seek to preserve cities in their historical roles." The commission even argued that policies to help declining cities were dangerous to the nation's economic health since to try "to restrict or reverse the processes of change—for whatever noble intentions—is to deny the benefits that the future may hold for us as a nation" (President's Commission, 1980, p. 4).

On the other side of this debate were those who attacked the abandonment of urban policy as "a form of Social Darwinism . . . with pernicious consequences, to individuals and social classes" (Barnekov, Rich, & Warren, 1981, p. 3). Much of Barry Bluestone and Bennett Harrison's book, *The Deindustrialization of America,* chronicled the devastating social impact on local communities when plants shut down and move their operations elsewhere. In the face of mounting evidence that entire regions and metropolitan areas were in the midst of a deindustrialization process, scholars and political figures in the U.S. called for a new round of national urban and regional policy. The National Development Bank proposed by President Carter in 1978 would have made federal dollars available to cities and areas with high unemployment caused by job losses.

Bitter political struggles were waged between the political leaders who favored such policies and Republicans and Sunbelt conservatives who helped put Ronald Reagan in the White House in 1980. It goes without saying that the conservatives subsequently succeeded in gutting national urban policy. The rout has not been as thorough in the parallel academic debate. For several years, for example, urban scholars built a minor industry analyzing and usually attacking Paul Peterson's assertion in *City Limits* that every city has a "unitary interest" in promoting economic growth and avoiding policies of redistribution (1981). The literature persuasively demonstrated that the benefits of economic development are usually narrowly distributed, and that "politics matters" in determining the nature of growth and its benefits (Stone & Sanders, 1988). It has become universally recognized that the recent economic restructuring in cities is accompanied by increasing income inequality and the proliferation of social problems (Squires, 1989).

Like the policy debate that it parallels, the scholarly debate has been overtaken by events. A significant number of older cities in North America and Europe arrested their slide in the 1980s and began to experience a degree of economic recovery. The new questions deal less with whether older cities will or should survive, and more with whether and under what circumstances cities can adapt to a changing world economy and a new international urban hierarchy. As cities in all the advanced Western nations adjust in different ways to the global economy of the 1980s and 1990s, the significance of local action finally has emerged as a relevant, even compelling topic for analysis.

During the 1980s studies of global restructuring proliferated. For the most part, cities were described as helpless pawns of international corporate elites. These studies described the systemic forces in the international economy that were forcing older industrial cities and regions into decline. Little significance was attached to the role of human agency in accelerating, retarding, or reversing the process of economic change (Smith, 1988). To some degree, there was an ideological resistance to such a focus, insofar as many scholars, and not only those on the left, were reluctant to give up the idea that political movements might still force a new round of economic and urban policies from national governments. A focus on actions at the city level required, in effect, an admission that cities would have to adjust in some way, or face the prospect of slipping into irreversible decline. In the end scholarly interests fell in line with changes in the real world.

As we have argued elsewhere, "most cities are not the helpless pawns of international finance, industry and commerce. They are in a position to mediate and direct their own destinies" (Parkinson & Judd, 1988, p. 2). In this context it is not merely a habit of language to speak of cities as organic entities. Though cities are made up of a multitude of social, economic, and political groups and factions, urban elites generally assume the role of trying to project an image to the external world of a city with a distinct history and identity. Politics certainly matters, even when the "city as a whole" is the unit of analysis: One of the most important outcomes of local politics is the emergence of an elite constellation or coalition that can speak for a city. This is the essence of Gramsci's concept of hegemony—the capacity to "guide" (Gramsci, 1971). When a local elite achieves this capacity, the "accidental city"—one whose destiny has been "dictated by locational advantages, technology and market forces"—may be transformed into an "intentional city" (Knight, 1989, p. 225-226).

GLOBAL RESTRUCTURING AND THE IMPACT ON CITIES

To understand the pressures on individual cities to revitalize their local economies, it is essential to appreciate the scale and pace of the restructuring of the international economy. During the 1950s and 1960s, international trade, capital investment, and labor migration patterns contributed to rapid economic growth in the Western industrial nations. Capital-intensive exports were sent from these countries to nations at the periphery of the integrated financial and trade system that tied the core nations to one another. Low-cost energy was imported from several key oil-producing countries, and basic raw materials were extracted from the so-called "developing" nations (Glickman, 1989, p. 69). But in the 1970s these relationships began to unravel. Oil prices ratcheted up in 1973 and for several years after, but in the longer term, the old arrangements were destabilized by the globalization of investment and production. Capital investment and manufacturing operations began to flow to the developing countries as multinational corporations sought to drive down their costs of production by moving assembly operations to places with low labor.

The expansion of world trade and international investment flows was facilitated by and also provoked the internationalization of business and finance. Global corporations developed the ability to shift productive enterprises and investment capital rapidly from place to place. Conglomerate

TABLE 1.1
Structure of Civilian Employment in Selected Advanced Capitalist Countries, 1960–1981, Illustrating the Rise of the Service Economy

	Percentage of Employed Population in:								
	Agriculture			*Industry*			*Services*		
	1960	*1973*	*1981*	*1960*	*1973*	*1981*	*1960*	*1973*	*1981*
Canada	13.3	6.5	5.5	33.2	30.6	28.3	53.5	62.8	66.2
France	22.4	11.4	8.6	37.8	39.7	35.2	39.8	48.9	56.2
W. Germany	14.0	7.5	5.9	48.8	47.5	44.1	37.3	45.0	49.9
UK	4.1	2.9	2.8	48.8	42.6	36.3	47.0	54.5	60.9
USA	8.3	4.2	3.5	33.6	33.2	30.1	58.1	62.6	66.4
OECD	21.7	12.1	10.0	35.3	36.4	33.7	43.0	51.5	56.3

SOURCE: OECD Labour Force Statistics. Adapted from David Harvey (1989, p. 157).

corporations mushroomed in size and administrative capacity through expansion to new locations and the acquisition of companies producing different and diversified products and services. The task of corporate managers was to constantly calculate profit margins, cost efficiencies, transportation advantages, labor market characteristics, tax levels, and national policies. Multinational corporations ceased to identify with any particular place or even with a particular nation (Barnet, 1974). The "efficiency" of one location over another became the touchstone for corporate investment decisions. And as a consequence, patterns of investment within the industrial nations changed very rapidly. The semiconductor, textile, clothing, and shoe industries relocated to low-labor sites; auto, chemical, steel, rubber, and other heavy industries dispersed their operations around the globe. New plants using more efficient technologies were built in regions within the core countries or in the developing countries in direct competition with the plants located in the older industrial regions. The older regions slid into rapid, often abrupt decline or collapse.

As shown in Table 1.1, between 1960 and 1973 industrial employment fell in the U.S., Canada, the U.K., and West Germany. Only in France did it increase as a proportion of the labor force. Service-sector employment rose substantially in all five countries, and agricultural employment continued its long-term decline. From 1973 to 1981, a period that encompasses the recession of the 1970s that hit all the advanced industrialized nations, industrial employment went into a nosedive. Statistically, it was compensated for by gains in services jobs, but those jobs were not necessarily, or even generally, located in the same regions or metropolitan areas that suffered sharp declines in manufacturing. Indeed, service-sector growth tended to take place disproportionately in new regions—the Sunbelt in the U.S., the Toronto area in Canada, the "sunrise" region in the south of England, Paris and southern France, and southern West Germany.

Faced with the erosion of their most important economic sectors, old port and industrial cities and the regions that had been built upon a base of industrial production or resource extraction responded, initially, in a predictable way: Their elites sought favorable policies from the central state and, often simutaneously, central governments tried to shore up traditional economic sectors because the decline of these sectors was often interpreted as a portent of national economic crisis. This response was relatively short-lived in all the Western industrial democracies. When it became clear to urban leaders that they would not be bailed out by national policies, then their efforts—or the efforts of their successors—became redirected to finding ways to regenerate their local economies.

NATIONAL POLITICAL RESPONSES TO RESTRUCTURING

In the mid-1970s, a great deal of media attention in the U.S. was focused on the economic problems of Frostbelt metropolitan areas and the economic boom then underway in the Sunbelt. The term *Sunbelt* entered the popular lexicon following a major series published in *The New York Times* in 1976. Within weeks, coming hard on the heels of New York City's much-publicized fiscal crisis and with rising unemployment, urban and regional economic decline were suddenly hot topics. In a lead article titled "A Policy for Domestic Detente," *Business Week* suggested that the federal government should direct more aid to depressed areas of the country and equalize social welfare benefits among regions (Markusen, 1987, p. 162). The fact that a conservative business magazine made such a proposal demonstrated a widespread consensus that government had some responsibility to respond to uneven development, an assumption that had driven regional and urban economic development policies in the U.S. since the 1930s.

In other Western democracies it also was assumed that the central government should intervene to protect declining regions and cities. Indeed, in those countries, government policy was more far-reaching than in the U.S. For example, after World War II, Britain nationalized key industries and adopted planning legislation intended to guide the spatial location of industrial and residential development. France has long had a mixed economy, with substantial public control of larger enterprises. In reaction to economic difficulties in the 1970s, the state rapidly expanded into a wider range of industries on the premise that "nationalization appeared as the only alternative to 'multinationalization' (increased foreign penetration) and as a way to pursue extensive restructuring . . . " (Body-Gendrot, 1987, p. 240). And in West Germany, a "far-reaching and all-embracing restructuring and modernization process" was carried out between 1969 and 1982 that involved massive urban renewal projects in cities as well as state intervention to save manufacturing sectors (Mayer, 1987, p. 343).

In the United States, such a direct application of state power was ideologically impossible. Nevertheless, policies from the 1930s to the mid-1970s attempted to promote the economic health of declining cities and regions. Though the United States has never attempted to implement a national economic policy per se, many programs, including among others the Housing Act of 1937, the urban renewal and public housing program legislated in 1949, the Area Redevelopment Act of 1961, the War on Poverty and the Model Cities programs, and the Housing and Community Development Act of 1974, were enacted to revive local and regional economies. The urban wing of the Democratic Party was able to exact policies such as these from the national government for half a century, but in the late 1970s this era came to an end. In a bitterly contested congressional battle led by the Northeast-Midwest Congressional Coalition, amendments that favored older cities were added to the Housing and Community Development Act of 1978. But this was the last important victory for the urban wing of the New Deal coalition. With the election of Ronald Reagan in 1980, its influence evaporated.

As Ann Markusen has noted, the urban coalition could no longer speak powerfully for the nation's cities because cities were not all in the same condition, and they were not all represented by leaders with similar outlooks (Markusen, 1987, p. 160-189). A politics of economic decline and central city decay characterized the old urban areas of the Frostbelt. But in the Sunbelt, local economies were booming. And in any case, the leaders of Sunbelt cities were more likely to be drawn from the business community, to be Republicans not Democrats, and to favor laissez-faire policies rather than interventionist government. Subsequent attempts to expand national urban policy provoked intense regional antagonisms. The political outcome revealed how much in the space of a few decades the balance of power in U.S. national politics had changed.

Like the U.S., until the 1980s the Western industrial nations attempted to insulate their national economies and declining regions and localities from the effects of international restructuring. These policies took the form of reindustrialization or targeted aid to cities and regions. But in the first half of the 1980s an astonishingly swift policy revolution swept through the nations of North America and Western Europe. Adjustment rather than resistance to global economic change became the new *leitmotif* of national policies on both sides of the Atlantic.

Much of British urban policy that emerged after World War II assumed that public control over the physical environment and the provision of social benefits were necessary to overcome the legacy of unplanned and uneven growth inherited from the Victorian city. The idea that government should control the physical environment amounted to a demand for radical intervention by government—interventionism that would probably have been inconceivable before the war (Clawson & Hall, 1973, p. 38). In this period cities took on major roles managing housing, social services, and land use activities, which meant that although the central government provided a growing list of social benefits, government authority remained to some extent decentralized. Policy was designed to provide social and welfare support services to the victims of economic change in the cities more than to create wealth. And since disinvestment by the private sector was seen as the cause of many cities' economic decline, the public sector was regarded as the natural agency to lead urban reconstruction.

When the Conservatives took power in 1979, a major shift took place in British urban policy. The government defined the public sector, especially the "profligate" expenditure and "bureaucratic" planning by local authorities, as the cause of inner cities' problems and the private sector as the solution. Markets replaced politics as the primary response to urban decline; the values of urban entrepreneurialism replaced those of municipal provisions; private-sector leadership replaced public intervention; investment in physical capital displaced investment in social capital; wealth creation replaced the distribution of welfare; and a regime committed to centralizing public power weakened local autonomy.

The government decided that local authorities could not lead the economic regeneration of their cities and reduced many of their traditional powers and resources. A large number of initiatives were introduced to give the private sector the lead role in urban policy: city action teams, task forces, enterprise zones, free ports, urban development grants, urban regeneration grants, urban development corporations. The Conservative effort to restructure urban policy must be seen in a larger context, an attempt to change the ideological climate of Britain by creating an enterprise culture and replacing state action with market forces. The conservatives believed that economic growth could best be promoted if government stopped intervening to protect national, local, or regional economies from "inevitable" economic changes. Now it was possible that some cities, even regions, might not remain viable at all if they did not adjust to economic "realities."

In France, the long tradition of a centralized state had accustomed people to look to the state for leadership and initiative; " . . . for centuries, the French mentality had been accustomed to a strong state which claims to be the motor of improvement" (Body-Gendrot, 1987, p. 248). Accordingly, in reaction to the economic problems of the 1970s, the French government embarked on a program of nationalization and state support for key industries. The public sector grew very quickly, embracing such economic sectors as aircraft, arms, computer technology, automobiles, and steelmaking.

From 1971 to 1983 France engaged in a comprehensive program to modernize and defend French industry. The nationalization of basic industries increased public-sector employment from 20% of employees in firms of over 2,000 employees in 1971 to 48% in 1983 (Body-Gendrot, 1987, p. 241). However, the attempt to modernize ailing industries by the strategic application of state capital was very expensive. The state financed 225,000 jobs; funded job training schemes; issued work permits to 14,000 foreign migrants; and helped finance early retirement for 700,000 workers. Direct subsidies to industries doubled, and the budget for industrial research tripled. The basic philosophy guiding all this activity was "the view that there were no condemned sectors, only condemned technologies" (Body-Gendrot, 1987, p. 241). Thus, much of the money went to ailing sectors, such as steel.

By 1983 it became apparent that these policies were both expensive and ineffective. Unemployment began to rise, exports sagged, and imports increased. These events illustrated that in an integrated world economy, a single country could not go it alone and protect its national economic sectors from international pressures. Since 1983 policies have undergone a dramatic shift to encourage French industry to become more competitive. In their latest phase, these policies have involved "privatization, deregulation and tax incentives in a program which looks much like Thatcherism and Reaganism" (Body-Gendrot, 1987, p. 250). As applied to cities, this has translated into a program of decentralization. Cities are now expected to pursue economic development through policies initiated at the local level.

Much like France, West Germany responded to the economic problems of the 1970s by investing heavily in threatened industries. From 1969 to 1982, West Germany carried out a comprehensive policy to maintain the country's position as a leading exporter of high-quality manufactured goods. Though some industries were supported that did not ultimately maintain a competitive position, on the whole the policies succeeded in keeping West Germany competitive in the world economy (Mayer, 1987). But since 1982, West Germany has adopted policies of fiscal austerity and privatization. At the same time, the *Länd* and individual cities have initiated policies to "rationalize" land uses and to promote economic development. Political conflicts over economic development thus are played out in the cities (Mayer, 1987), and individual cities must try to create favorable environments for investment.

Canada is the one country represented in this book that has not had a coherent set of industrial or urban policies. One explanation is that Canada has few big cities, so that there has not been intense competition between them for favorable federal policy. Regional development has been construed as helping the provinces, as regions, through the redistribution of federal tax revenues to the poorer provinces such as Newfoundland. Special regional development programs have focused on rural areas.

In Canada, power is concentrated in the provinces, and cities are subject to provincial legislative and political control. However, the absence of direct federal intervention in city affairs does not mean that cities have not received federal assistance. Cities are linked to the federal government through Cabinet ministers representing their province. In this system, cities have successfully lobbied for important specific projects such as Montreal's Expo in 1967 and the expensive heavy rail system, which took people to the fair and now serves the city. But Canadian cities cannot rely on such grand projects for their economic health. As in other nations, they now must attempt to guide their future economic development.

THE ROLE OF LEADERSHIP

In this book we asked the authors to assess the role of leadership in achieving urban regeneration. We focus on three themes—capacity of leadership, choice of strategy, and the consequences of economic regeneration. Our first interest is in the capacity of leadership to implement an economic regeneration strategy in different cities and to identify some of the factors that contribute to the development of this capacity. We define leadership capacity primarily in institutional terms—the range, stability, and durability of local mechanisms and alliances which have been developed that allow a city to respond proactively to external economic pressures. A local leadership, among other things, can exploit its city's resources. These resources are diverse and differ from city to city, and would include the strength, structure, and stability of the private and public sectors in a city and the character of the political relations between the two. Important leadership resources also should include the locational or environmental assets of a city. A city's skill in exploiting national resources such as money, favorable legislation, and political support from government and national elites may be as important as the condition of the national economy.

If we speak of the capacity of cities to respond to external threats or opportunities, we actually are referring to the success of local elites in projecting a coherent interpretation of a city's "intentions" and of its economic and political environment—in other words, its "image." Political antagonisms within a city may be so great that no coherent response is possible, at least not one that seems representative of the city as a whole. However, a number of conditions may give rise to an elite with sufficient power to speak for a city. First and most obviously, an elite may dominate politics because it presides over a closed political system that gives little opportunity for opposition to arise. The classic party machines in U.S. urban history are examples of such systems, as are the business-dominated cities of the Sunbelt. Second, the formal concentration of legal or governmental authority may create a unified elite voice. Robert Moses, for instance, exercised almost dictatorial powers for more than 30 years in New York City by accumulating vast and overlapping formal authority (Caro, 1974). Finally, elites representing a city may have the capacity to speak with a coherent voice because the political process facilitates bargaining, negotiation, and agreement among a broad range of political and social groups.

Even when elites wish to respond proactively, or when they seem to have a mandate to do so, they may be limited by their own vision or ideology. In many U.S. cities, for example, governing elites often are persuaded that their only option is to create a "good business climate" by minimizing government, taxes, and regulation (Stone & Sanders, 1987). In the U.K. or continental Europe, the opposite example might not be hard to find of local elites that have regarded the private sector with suspicion or indifference.

Of course even well-organized, visionary urban elites are limited to options that are realistically available. Indeed, the ability to assess options is itself a measure of the capacity of a local elite. Every city has its own unique problems and opportunities. The constraints and opportunities facing a city interact in a complex way with the character of its politics and the capacity of its leadership. Harvey Molotch, for example, has argued that

> while virtually all localities may be run by growth machine elites, there are differences in the quality and quantity of growth that each urban area can plausibly attract. If this were not the case, we would have to argue that the relative size of cities is simply a result of the energy and cunning of their property entrepreneurs vis-à-vis their counterparts elsewhere [Molotch, 1988, p. 29].

The energy and cunning of local elites matter a great deal because the search for options can be done ineptly. One example is the building of the Renaissance Center in Detroit, which stands as a monument to corporate power in the midst of urban decay (Molotch, 1988, p. 34). To avoid this kind of mistake, "Growth elites must squarely face the constraints that nature and the world economy hand them, but then move heaven and earth to maximize the possibilities that are nevertheless possible . . . " (Molotch, 1988, p. 35).

The authors of the case studies in this book describe how urban leaders responded to economic crisis, and also assess the consequences of the economic regeneration that has taken place. The 12 cities are: Baltimore, Buffalo, Houston, and Pittsburgh in the United States; Montreal and Vancouver in Canada; Glasgow, Liverpool, and Sheffield in the United Kingdom; Marseilles and Rennes in France; and Hamburg, West Germany. All of these cities have been substantially affected by the restructuring of the global economy and the impact of this restructuring on the national economies in North America and Europe, and most of them have produced or are attempting to produce a leadership or political coalition around the issue of regeneration.

The cities are second tier metropolitan centers that were historically important as port, commercial, or industrial centers. They have not evolved into world cities like New York, London, Paris, Tokyo, Hong Kong, and San Francisco, which contain dense clusters of corporate headquarters, banking, corporate services such as insurance, accountancy and law, and research and government (Friedmann & Wolff, 1982). To a considerable extent, the cities studied in this volume are dependent on investment and finance decisions made by the institutions concentrated in the world cities.

Our case studies reveal considerable variation in the ideologies embraced by local elites, the degree of elite coherence, and degree of coordination among elite factions and political groups, and they describe a wide variety of strategies to regenerate local economies. But these differences are bounded by this basic fact: In all of the cities, the goal is to "establish and maintain an economic climate in which the private developer is willing to take risks" (Levine, 1989, p. 22). The case studies show that there are a variety of ways that this goal may be pursued.

TWELVE CITIES

The four case studies from the United States profile the experiences of Baltimore, Buffalo, Houston, and Pittsburgh. Baltimore is one of the original Colonial seaport cities, and in the nineteenth century it became a significant manufacturing center specializing in shipbuilding, primarily metals, and steel production. Though as recently as the 1940s it was ranked as the seventh-largest manufacturing center in the country, it did not historically attract corporate headquarters or management. As a grimy seaport and industrial city, it ". . . had no obvious attractions; it was a city to be avoided rather than visited" (Law et al., 1988, p. 60). But by the 1980s this reputation had changed fundamentally. The spectacular Inner Harbor development draws thousands of tourists each year, giving part of the central city a prosperous look even though, Richard Hula notes in this volume, large areas of dilapidated housing and areas of poverty and social decay exist just a few blocks away. Hula explores the argument that the lack of spillover benefits to the rest of Baltimore can be traced to the fact that a virtual "shadow government" of quasi-private redevelopment agencies have presided over the harbor's redevelopment. These agencies operate outside of public control or even public scrutiny, and, as a consequence, the decisions concerning the direction of economic regeneration have been made behind closed doors.

Buffalo is located in western New York state, in a region that was once an industrial giant in the U.S. economy. Its economy has now collapsed so completely that it has not even been able to participate in a process of economic restructuring; instead, it is entirely "dependent on outside market and state forces for its economic and social renewal" (Perry, 1987, pp. 113-114). In his study, David Perry describes how Buffalo's leadership was shattered along with its economy. Recent efforts to revitalize Buffalo have been focused on attempts to hire talented professionals from outside the city to strengthen public sector agencies enough to build a new leadership base.

Since the 1930s Houston had been a center of the world oil industry and until the 1980s it was the archetype of the booming Sunbelt city (Hill & Feagin, 1987, p. 155). Rising oil prices that adversely affected other regions and cities in the 1970s redounded to Houston's benefit. For instance, in a press conference held in October 1981, President Ronald Reagan held up a Houston newspaper to show off a long list of job ads and recommended that people living in cities where unemployment was high should "vote with their feet" and move to the more prosperous areas of the country (*New York Times*, 1981). Ironically, only a few months later Houston's economy went into a nosedive when world oil prices collapsed. Reflecting the Sunbelt's traditional notions of entrepreneurialism, the efforts in the 1980s described by Robert Parker and Joe Feagin to revive the local economy have rested on the assumption that a "good business climate" is one with low taxes, few environmental regulations, and a weak public sector.

If Houston is the prototypical Sunbelt city of the 1970s, Pittsburgh has served as a symbol both of Frostbelt decline and of skilled leadership in the cause of regeneration (Sbragia, 1989, p. 104). While its region's economy reeled from the loss of manufacturing jobs, between 1950 and 1980, the proportion of the city's manufacturing fell from 26% to less than 15%. The city lost almost one-quarter of its population between 1970 and 1984 (Sbragia, 1989, p. 105). The city's ability to cope with rapid economic change can be traced to aggressive economic development and renewal projects that were launched as early as 1946, making Pittsburgh the nation's first city to construct a well-organized coalition behind renewal. In this volume, Alberta Sbragia describes both the history of these efforts and Pittsburgh's unique model of development, which involves aggressive governmental leadership working in a coalition with nonprofit cultural, medical, and educational institutions.

The three British cities, Glasgow, Sheffield, and Liverpool, are Labour-controlled cities that have lost their historic port and manufacturing functions, and all were deeply affected by the recession of the 1980s. Glasgow's economy began to falter shortly after World War I, in step with a decline in shipbuilding. But World War II and the boom that followed it during the period of European reconstruction brought a last surge of prosperity. The marked decline that began in the late 1950s changed quickly to full-scale depression in the local economy in the 1970s. By 1976, when the Scottish Development Agency was created to regenerate the region's housing and economy, unemployment rates in the Glasgow area exceeded 20%. As a result of its efforts, by the 1980s Glasgow has received huge amounts of public funds, and its inner-city area has been substantially renewed. The infusion of public funds has succeeded in leveraging private-sector investment and a surge in service-sector employment.

Liverpool's economic decline parallels Glasgow's. But Liverpool tried to solve its problems in the 1980s through a unique strategy of confrontation with the central government. As Michael Parkinson's study shows, this course of action proved to be disastrous. In an attempt to force Margaret Thatcher's Conservative government to give more money to the city, for several years the Labour council threatened to bankrupt the city as part of a strategy to provoke a political crisis that might be a building block for a national coalition uniting Labour-controlled cities. When the councillors were prosecuted and about half the council was disqualified from holding office, this strategy fell apart (Parkinson, 1985). In the aftermath, it has been difficult for Liverpool to forge a coalition that can lead a well-organized program of economic revival.

In the late 1970s, as a severe recession hit the British steelmaking industry, Sheffield, which for more than a century had built its local economy on steel and heavy engineering, was plunged into a crisis. Rather than confronting the central government, its Labour party responded with a version of local socialism, which claimed "for a locally elected authority like Sheffield the right for greater community and workers control and influence over employment and the local economy" (Lawless, chap. 7 of this volume). This "radical" era emphasized public sector intervention to try to create new jobs; equity financing for small firms; and efforts to promote equal opportunity employment. By 1986 these programs of municipal socialism gave way to an emphasis on cooperation between local government and private employers, the external promotion of the city as a "good business environment," and support for flagship projects in retailing, leisure, and commercial development. Paul Lawless charts this transition to market-oriented strategies and the emerging belief that Sheffield's prosperity ultimately depends on creating a new economic base that relies less on manufacturing and more on new service-sector industries funded by outside investors.

Montreal and Vancouver both built thriving economies when they evolved as major entrepóts along the east-west axis of Canadian national development. As a center of trade and commerce linking Canada and Britain, Montreal became a leading port city; by 1929, it was the most important port for grain exports in North America. In the ensuing decades it also became a major industrial and manufacturing center. But beginning in the late 1960s, its port and manufacturing base eroded quickly. In addition to the impact of the worldwide recession of the 1970s, Montreal was hit hard by the rapid integration of the Canadian and U.S. economies, which created north-south economic linkages, with Toronto as the chief beneficiary. The decline of Montreal's traditional economic role has provoked attempts to transform the city's economy—while preserving its reputation and identity as an international city. In the 1960s and 1970s "grand projects" emphasized its international and cultural identity—most dramatically, the 1967 World's Fair and the 1976 Olympic Games. In the late 1980s, as Léveillée and Whelan show, the city's leaders have concentrated their economic development efforts on infrastructure and activities to promote the city as a location for international finance and corporate organizations, international conferences, and tourism.

Vancouver evolved as a major entrepót linking Canada to the Pacific Rim. Its industrial base was never well developed; rather its economy was built around the trans-shipment of agricultural goods and timber from the vast developing region of western Canada through its port. But both agriculture and the forest industry have been adversely affected by falling commodity prices since the 1970s, which affected the economy of the province even more than Vancouver's. In the 1980s the city government, behind the leadership of an aggressive mayor, began to map out a strategy to promote Vancouver as the communications, financial, services, and tourism center linking Canada to the Pacific Rim. Initially, the plan called for policies requiring the city to coordinate the business community's economic development efforts and invest in new infrastructure to support a rapid increase in office construction and tourism. At the same time, the plans called for maintaining the city's "quality of life." But since 1986, with the election of a conservative mayor and council, this strategy has given way to one that promotes Vancouver's business climate to outside investors through land-use deregulation and passive government.

In continental Europe, as in North America, urban elites also have emerged around issues of economic regeneration. Though like most European cities Rennes did not go through a crisis of its inner-city areas, in the 1970s the local industrial sector did begin to seriously erode. After some searching for solutions, in the 1980s an economic development strategy emerged in which banks and administrative agencies of the centralized state, together with a local intellectual elite united within a political party, organized a broad coalition of interests behind a comprehensive plan to transform Rennes into a significant European metropolis. The French tradition of state-supported public planning, the existence of a strong central state that has historically intervened in or even controlled local affairs, and, in the case of Rennes, the presence of educational, cultural, and religious elites, came together to facilitate a strategy for economic development that united the various groups and sectors in the city. Strong local leadership has been a key element holding this complex process together.

As the first port and second city of France, Marseilles benefitted greatly from central state policies from the 1950s to the early 1970s and became the site of massive industrial and urban development programs. Marseilles undertook its own program of infrastructure and public services development to support the state-financed projects. But all of these policies could not change a basic fact: The port of Marseilles always has been sensitive to changes in the international economy. The fluctuations in the oil market that began with the oil embargo of 1973 provoked an economic crisis that laid bare two little-understood realities: Much of the port activity already had moved outside of Marseilles to nearby locations; and the industrial base in the city itself also had decentralized or was in the midst of decline. After the collapse of the city's political machine in the 1980s, local political elites were not able to respond effectively to these developments. The warring factions that constitute local politics in the city must reach some accommodations if Marseilles is to fashion strategies to regenerate its economy.

West German cities possess very significant powers of their own, which makes Hamburg a good study of the impact of a decentralized system that grants local political elites power and resources to implement economic regeneration strategies. Until recently the ancient port city of Hamburg was "one of the most prosperous urban areas in Western Europe" (Law et al.,

1988, p. 83). It is the main commercial center of northern Germany, with an economy historically based on the port, port industries, and international trade. Hamburg participated in the rapid economic development of Western Europe that occurred in the postwar years. However, its economy weakened noticeably in the 1970s when the slow growth of exports in the world economy undermined its export-based economic structure. The responses of political elites to these changes have been far-reaching. Initially, policies were implemented to support traditional sectors of the economy. In the 1980s these have given way to policies that target more profitable activities in the service sector and urban renewal schemes to "gentrify" the city and make it more attractive to investors and visitors.

DIFFERENCES OR CONVERGENCE?

Our case study cities are located in five countries with different ideological, institutional, political, cultural, and economic traditions. The five countries vary in governmental structure and degree of centralization, the scale and composition of their private sectors, the ideological structure and significance of their party systems, and the degree of legal, political, and fiscal autonomy for cities and regions. However, despite these important differences, all five countries, and their cities, have been subject to similar economic pressures over the past decade.

In her article on the changing world economy and urban restructuring Susan Fainstein provides a context for understanding how cities have responded to economic decline. On the one hand, if we assume a global perspective, cities in all the advanced nations are in the position of adjusting to forces that they cannot control. The number of revitalization strategies cities can select from are quite limited, usually including variations on service-sector growth; thus, cities are converging, becoming more alike. But, as Fainstein points out, cities can be analyzed by looking at their particular history and social and political composition. When viewed from that perspective, the way cities respond to economic stress is determined significantly by their local characteristics. It is useful to note that the case studies in this volume are written from this vantage point.

There are systematic differences in how elites in the case study cities have responded to economic stress, and we have grouped the studies accordingly. In three of the cities, Pittsburgh, Rennes, and Hamburg (Part I), urban elites have developed a substantial capacity to define and implement coherent strategies to regenerate the local economy. Elites in these cities have forged broad-based coalitions to support regeneration efforts. A tradition of strong public leadership in each of these cities has sustained long-term strategies of regeneration.

The four cities grouped in Part II—Glasgow, Sheffield, Montreal, and Vancouver—have produced a leadership capable of guiding sustained regeneration efforts. However, the political coalitions necessary to back regeneration policies are fragile, or relatively new, and the policies adopted to restructure the local economy are as vulnerable and unproven as the political coalitions that give them support.

In two of the cities, Baltimore and Houston (Part III), regeneration efforts are controlled by private-sector elites that wield almost autonomous power. Municipal officials are marginal to the regeneration process, and the private-public partnerships that guide economic regeneration in these cities operate almost completely outside the public's view. This model of regeneration is probably typical of many cities in the U.S., and it is therefore important to understand how it operates, and with what effects.

The cities profiled in Part IV, Liverpool, Marseilles, and Buffalo, share the common characteristic of having serious leadership deficits. Elites in these cities have been unable to mobilize stable political coalitions, and as a result, they have not been able to implement coherent regeneration policies. The economic future of these cities may well depend on the emergence of new leadership.

Cities can, to some degree, determine their own futures. But what are their degrees of freedom? How well do local elites respond to external pressures? And what are the consequences of their actions? The case studies in this book map the terrain that will allow us to explore these questions further in the concluding essay.

REFERENCES

Barnekov, T., Rich, D., & Warren, R. (1981). The new privatism, federalism and the future of urban governance: National urban policy in the 1980s. *Journal of Urban Affairs, 3* (4), 1-15.

Barnet, R. (1974). *Global reach: The power of the multinational corporations.* New York: Simon & Schuster.

Bluestone, B., & Harrison, B. (1982). *The deindustrialization of America: Plant closings, community abandonment, and the dismantling of basic industry.* New York: Basic Books.

Body-Gendrot, S. (1987). Plant closures in socialist France. In M. P. Smith & J. R. Feagin (Eds.). *The capitalist city* (pp. 237-251). New York: Basil Blackwell.

Caro, R. A. (1974). *The power broker: Robert Moses and the fall of New York.* New York: Knopf.

Clawson, M., & Hall, P. (1973). *Planning and urban growth: An Anglo-American comparison.* Baltimore: Johns Hopkins.

Friedmann, J., & Wolff, G. (1982). World city formation: An agenda for research and action. *International Journal of Urban and Regional Research, 6* (309), 44.

Glickman, N. J. (1986). *The myth of the North American city.* Vancouver: University of British Columbia Press.

Gramsci, A. (1971). *The intellectuals* and *On education* (Reprinted). In Q. Hoare (Ed. and Trans.). *The modern prince and other writings.* New York: International Publishers.

Harvey, D. (1989). *The condition of postmodernity*. New York: Basil Blackwell.

Hill, R. C., & Feagin, J. R. (1987). *Detroit and Houston: Two cities in global perspective*. Cambridge, MA: Basil Blackwell.

Knight, R. V. (1989). City development and urbanization: Building the knowledge-based city. In R. V. Knight & G. Gappert (Eds.). *Urban affairs annual reviews: Vol. 35. Cities in a global society* (pp. 223-244). Newbury Park, CA: Sage.

Law, C. M., Grune, E. K., Grundy, C. J., Senior, M. L., & Tuppen, J. N. (1988). *The uncertain future of the urban core*. London and New York: Routledge.

Levine, M. V. (1989). The politics of partnership: Urban redevelopment since 1945. In G. S. Squires (Ed.). *Unequal partnerships* (pp. 12-34). New Brunswick: Rutgers University Press.

Markusen, A. R. (1987). *Regions: The economics and politics of territory*. Totowa, NJ: Bowman & Littlefield.

Mayer, M. (1987). Restructuring and popular opposition in West German cities. In M. P. Smith & J. R. Feagin (Eds.). *The capitalist city* (pp. 343-363). New York: Basil Blackwell.

Molotch, H. (1988). Strategies and constraints of growth elites. In S. Cummings (Ed.). *Business elites & urban development* (pp. 25-48). Albany: State University of New York Press.

The New York Times. (1981, October 23).

Parkinson, M. (1985). *Liverpool on the brink*. Berks, England: Policy Journals.

Parkinson, M., & Judd, D. (1988). Urban revitalization in America and the U.K.—the politics of uneven development. In M. Parkinson, B. Foley, & D. Judd (Eds.). *Regenerating the cities: The UK crisis and the US Experience* (pp. 1-8). Manchester, England: Manchester University Press.

Perry, D. C. (1987). The politics of dependency in deindustrializing America: The case of Buffalo, New York. In M. P. Smith & J. R. Feagin (Eds.). *The capitalist city* (pp. 113-137). New York: Basil Blackwell.

Peterson, P. (1981). *City limits*. Chicago: University of Chicago Press.

President's Commission for a National Agenda for the Eighties. (1980). *Urban America in the eighties: Perspectives and prospects*. Washington, DC: U.S. Government Printing Office.

Sbragia, A. (1989). The Pittsburgh model for economic development: Partnership, responsiveness, and indifference. In G. D. Squires (Ed.). *Unequal partnerships* (pp. 103-120). New York: Basil Blackwell.

Squires, G. D. (Ed.). (1989). *Unequal partnerships*. New Brunswick, NJ: Rutgers University Press.

Stone, C. N., & Sanders, H. T. (Eds.). (1987). *The politics of urban development*. Lawrence: University of Kansas Press.

2

The Changing World Economy and Urban Restructuring

SUSAN S. FAINSTEIN

RESTRUCTURING AND LOCALITY

There are two ways of analyzing cities, neither incorrect. The first, or global, approach scrutinizes the international system of cities (and its national and regional subsystems). While noting particularities, this mode of explanation attributes them to the niche or specific node that a city occupies within the overall network. Scholars using this perspective predict uneven development and consequent territorial difference; from their vantage point, which particular places win or lose matters less than that there inevitably will be winners and losers.[1] In contrast, the second approach, which works from the inside out, examines the forces creating the particularities of a specific place—its economic base, its social divisions, its constellation of political interests, and the actions of participants. Within the first framework, differences among cities are manifestations of the varying components that comprise the whole. The second traces urban diversity to internal forces and the tactics used by local actors.

The same city can thus be regarded both as part of a totality and as a unique outcome of its particular history. To offer an example that illustrates the point: It is possible to tell the story of Houston using either analytic framework (see Parker and Feagin in this volume). Using a world system approach, we see Houston as building its prosperity on its unique function as the center of the U.S. oil industry and the headquarters of firms dominating world petroleum exploration and marketing. The economic decline that it suffered during the eighties resulted from global economic factors including plummeting oil prices and overvaluation of the dollar, which heavily damaged the ability of Texas manufacturing firms to export. Moreover, its role as a regional financial center weakened as a consequence of bank failures caused by over-extension during the preceding boom period, particularly lavish financing of real estate development, which itself was premised on ever-increasing affluence. From this perspective Houston's rise and decline can be traced to its place in controlling, financing, and marketing one of the most important commodities in international trade, and one that has been particularly affected by world political and economic currents.[2]

Most important, this approach provides insight into the general relationship between macroeconomic forces and urban outcomes. The argument is as follows: Changing modes of corporate finance and control, causing and produced by the geographic decentralization of production, globalization of financial and product markets, and internationalization of the giant corporation, increase the vulnerability of places to disruptions in the markets of commodities on which they are dependent for their economic well-being. Moreover, the instability of foreign exchange levels increases their exposure to uncontrollable outside forces, regardless of their efficiency of production, since it causes the world-market price of their output to vary independently of their production costs.

The inside-out approach to explaining urban restructuring, on the other hand, allows us to identify the dynamic factors driving Houston's adaptation to changing circumstances. Applying this analytic mode, we explain the city's past status by relating its development as the capital of the petroleum industry to its entrepreneurial culture and favorable business climate. We identify the industry leaders who founded enterprises in Houston and trace the city's expansion to federal subsidies attracted by well-connected politicians (Feagin, 1988, chap. 6). One reason for the sharpness of Houston's recent decline and the extremity of its effects on its poorest residents is the past reluctance of the public sector to intervene in the economy and to provide social welfare. At least part of the explanation for Houston's present turnaround lies in an increased willingness to plan and manage growth.

We can similarly assess the decline of many other old commercial and industrial cities and the regeneration of some. Probably no city in the advanced capitalist world has been unaffected by the reorganization of the global economy of the past two decades. For those places especially dependent on dying manufacturing and port industries, local leadership has been one element that could improve their competitive position, although always within the serious constraints posed by historic economic base, regional location, and national policy.

To understand the process by which improvement has occurred, we must identify the changes in economic functions resulting from shifts in the world and national economy and examine the activities of groups and leaders within particular cities that affected their new roles. Thus, Pittsburgh and Sheffield similarly suffered from the world's increase in steel capacity, reduced demand for metals products, and heightened international competition. Both have had local leadership that sought to restructure their economies so as to develop new economic functions and indeed have seen economic revivals. A similar transformation has occurred in the old port cities of Baltimore and Hamburg: After long periods of decline, enterprising municipal administrations identified new opportunities and managed to attract outside investment to their locales. In the case of Baltimore these were primarily tourist oriented, while Hamburg followed the high-tech route.

This chapter briefly recapitulates the now extremely familiar story of economic restructuring and urban transformation. It then examines the varying interpretations of urban trajectories and potential according to the two vantage points sketched above. It will summarize the right and left ideological responses to economic restructuring; and finally it will set forth the types of policies available to progressive local regimes and oppositional movements, concluding with a discussion of the relationship between the politics of locality, economic forces, and national governments.

ECONOMIC CHANGE AND URBAN RESTRUCTURING

We remember now only with difficulty the immediate post-World War II period, when industrialization seemingly offered the key to economic prosperity. Cities with large, diverse manufacturing bases promised secure growth and stable employment; the Soviet Union set its goal as outpacing the West through the development of heavy industry; and the task in front of war-ravaged Europe was the reconstruction of its manufacturing capacity. Now, in a world awash in commodities, peripheral locations have become the most advantageous sites for manufacturing. The future of older cities appears to depend on capturing the financial, informational, and managerial functions that determine the world's capital flows, although some areas can alternatively rely on tourism, scientific or medical services, and high-technology manufacturing to maintain a competitive edge. Overall, in the advanced capitalist world, massive employment losses in manufacturing sectors have been balanced or mitigated by gains in services and wholesale and retail trade; many places, however, have never fully recovered from the rapid loss of manufacturing jobs and are still characterized by high unemployment rates and continued outflow of population.

One of the main lessons of the past two decades is that the economic composition of places seems to have become less and less permanent (see

Harvey, 1989). While restructuring of manufacturing industry may have passed its peak, a similar rationalization of tertiary industries has possibly just begun. During the 1980s many cities have shown signs of regeneration, evidenced in new office towers, gentrified inner-city neighborhoods, and job creation. However, the internationalization of economic competition, which was one of the principal causes of manufacturing decline, also threatens this new vitality. While globalization has enhanced the importance of financing, informational, and control functions, it has also enlarged the number of competitors in the tertiary sector. Within Europe each nation has hopes of housing the control center of the European Community after 1992 and is competing fiercely to attract headquarters and financing operations; the glut of office space that currently characterizes United States metropolitan areas may soon spread to Europe, where numerous large office projects are under construction in anticipation of European union. In the United States financial interests look warily at the expanding Japanese presence in the banking and investment industries, which threatens to make Tokyo the world's financial capital.

Within each country, even domestically owned financial and service firms have become increasingly footloose as they emulate industrial corporations by separating their routine processing functions from more complex operations and decentralizing them to low-cost areas. Furthermore, prosperity based on the advanced-service and financial sectors remains hostage to the health of financial markets. The shock of October 1987 continues to reverberate through diminished employment in financial sector firms and reduced consumption in cities dependent on that industry.

Even successful regeneration, therefore, demonstrates signs of instability and social fragmentation. While financial centers benefit from merger and acquisition activities, other cities find that consolidation of the new conglomerates results in the closing down of formerly profitable establishments now redundant or too encumbered with debt to remain viable. Efforts to spur central business district development and the "realistic" dismissal of manufacturing as the future basis for growth have displaced residents and small firms and left blue-collar workers stranded.

Along with deindustrialization has come the decline of a homogeneous, relatively well-paid working class and the growth in size of income strata at both extremes of the spectrum (Harrison & Bluestone, 1988, chap. 5). The outmigration or closing of factories and obsolete shipping facilities has produced desolate landscapes of unused structures in once central locations. Changes in social groupings have resulted in homelessness and the decay of formerly stable working-class residential districts, on the one hand; on the other, they have heightened the demand for converted, well-located structures and luxury new construction. Combined with demographic changes due to dropping birth rates, growing numbers of single-member and female-headed households as well as high-income, two-earner couples, and large-scale immigration, these factors have heightened the fragmentation of urban space (see Mingione, forthcoming). Moreover, cities with low unemployment rates and new investment still face fiscal problems that severely restrict governmental efforts on behalf of low-income residents and limit necessary investment in the physical infrastructure required for future growth.

Social scientists generally agree that the trickle-down effects of new development in "successfully" restructured cities have excluded a large proportion of the population and may even have worsened their situation (Parkinson & Judd, 1988; Squires, 1989). For the many locales that remain trapped in the trajectory of industrial decline and high unemployment, circumstances are obviously worse. To formulate a political stance that effectively addresses the economic distress of old cities requires identifying points of indeterminacy as global forces operate on particular places. Only after such an analysis do local policies stand a chance of stimulating growth; it, however, offers no guarantee and in some cases may prove extremely discouraging to local action.

THE GLOBAL PERSPECTIVE

Localities are forever in the position of adjusting to forces beyond their control. The oil crises of the 1970s, the rise of the manufacturing economies of the Far East, the management failures of Western oligopolistic industries, the rationalization of firms through decentralization of their various components into least-cost locations, global sourcing, and modern telecommunications have all had profound effects on urban economic structures. While technology is not the cause of increased capital mobility, the loosening of natural and technical constraints on location has allowed firms to further exploit socially created locational advantages (Fainstein & Fainstein, 1988). Most important of these advantages is the group of attributes often called the business climate. Also significant is proximity to markets, which matters far more to most businesses than closeness to raw materials or natural features like rivers.[3] Proximity to markets, however, may only require location close to an airport rather than placement within an actual agglomeration.

For many medium-size cities the weakening in importance of their natural advantages has meant the termination of their *raison d'etre*. For example, although port facilities continue to be important generators of economic growth, their existence depends less on the quality of the available water berths than on the pricing of dock labor and the presence of modern container-handling operations. The enhanced capacity of a few ports to handle greatly increased amounts of tonnage along with the ease of transferring containerized loads to trucks and trains reduces the need for numerous ports dotting a single shoreline (Hoyle, 1988). While for London and New York the decline of the port was within the context of diverse other economic functions, for Liverpool or Baltimore port-related activities defined their specialized niche in the system of cities. Consequently, they were particularly vulnerable to technological transformation, and they lacked economic leaders capable of developing other functions since they had few sectors independent of the port. A similar problem exists for the steel-fabricating areas of northern France or the American Midwest.

Such cities then are systemically disadvantaged. The lack of vital private business outside the declining sectors leaves only the public sector to offer a potential engine for stimulating new growth. For the public sector to do so means finding a new niche that the locality can occupy.[4] It is in this possibility of identifying a new niche that indeterminacy exists within the global framework of analysis, and it is here that the two approaches to urban analysis complement each other (see Riley & Shurmer-Smith, 1988).

Urban growth coalitions, however, find themselves in a prisoner's dilemma in that their success in finding a new area of specialization depends on leadership groups elsewhere not initiating the same strategy. Festive retailing may work for the first cities that revitalize their waterfronts using this formula, but impulse-buying tourists can support only so many stores selling brass ships' furnishings and Irish shawls. Research parks have spurred development in Cambridge, England, and Charlotte, North Carolina, but they are predestined to languish in most places. Just as in the market economy as a whole, latecomers to an industrial sector will not see the profits of the innovators, so cities that are imitators are unlikely to flourish. Hence, while a city's economic leadership has leeway in choosing new niches, it does so within the framework of a system of competing cities, putting the public's investment at risk.

THE LOCAL AUTONOMY PERSPECTIVE

Viewing from the inside, we can see each city as having a potential for regeneration that is dependent on the actions of its constituent groups. There are three major dimensions on which cities vary according to the character of state intervention aimed at economic regeneration: (1) extent of governmental entrepreneurship, (2) amount of planning, and (3) level of priority to those in greatest need. Which growth strategy is followed within a city and the city's commitment to targeting low-income groups are consequences of political struggle and are largely independent of external forces (Smith, 1988)—although, as indicated above, whether the growth strategy works is less open.[5]

ENTREPRENEURSHIP

Whereas city governments once restricted their activities to building infrastructure and providing services, virtually all now take an active role in promoting economic growth. Eisinger (1988) contends that, within the U.S., city and state governments have moved from an initial, naive "supply-side" strategy for stimulating private investment to "demand-side" policies:

> What guides the entrepreneurial state is attention to the demand side of the economic growth equation. Underlying the actions of the entrepreneurial state is the assumption that growth comes from exploiting new or expanding markets. The state role is to identify, evaluate, anticipate, and even help to develop and create these markets for private producers to exploit, aided if necessary by government as subsidizer or coinvestor. The policies of the entrepreneurial state are geared to these functions. They include the generation of venture capital for selected new and growing businesses, the encouragement of high-technology research and product development to respond to emerging markets, and the promotion of export goods produced by local businesses to capitalize upon new sources of demand [Eisinger, 1988, p. 9].

This terminology is confusing since the term "demand side" usually refers to a policy that subsidizes consumers rather than investors. By *demand side* Eisinger simply means a more entrepreneurial or active policy that identifies market opportunities rather than indiscriminately subsidizing all investors. His general point, however, is that subnational governments in the United States increasingly seek to encourage specialized development where their economic policymakers have identified a strategic advantage. He considers that this entrepreneurship represents a new and important role for the subnational state, although he considers the resultant programs of "modest dimensions and uncertain impact" (Eisinger, 1988, p. 34).[6]

PLANNING

Cities also vary according to the amount of planning they do, both within countries and from country to country. The United States and Great Britain differ considerably from Canada and continental Europe in the extent to which growth is channeled through the planning process. The construction of La Defense in Paris as a corporate center, for example, and the current development of that city's southeast sector as a financial district result from a very strong governmental role in guiding development (Savitch, 1988, chap. 5). In contrast, almost all U.S. cities allow office developers broad limits within which to choose their sites. British restrictions on office locations have been greatly relaxed, allowing simultaneous competing projects. According to the head of the London Regional Planning Office, "boroughs that had formerly tried to stop office development will do so no longer." Or, in the words of the chief planner of one of London's boroughs, "even developers would like more planning."[7]

Growth regulation for environmental protection, the preservation of low-income housing, or the maintenance of manufacturing sites is frequently criticized. Critics assert that such planning is, depending on whether it is in growing or declining areas, either exclusionary or a luxury that deteriorating communities cannot afford (Sternlieb, 1986). Without planning, however, urban landscapes become the product of impersonal market forces, dominated by the interests of capital (Foglesong, 1986). Not only does the absence of planning prevent the general public from being able to affect urban outcomes, but it also denies real-estate interests a regulatory body to insure against overdevelopment. Consequently we see the oversupply of office space that now threatens the future stability of regenerating cities.

ECONOMIC GROWTH AT THE BOTTOM

Within the capitalist countries that have undergone restructuring, urban regeneration has largely taken place under elite leadership, although some exceptions like the "third Italy" exist. As might be expected from a top-down phenomenon, participating economic elites have been the primary direct beneficiaries of the growth of new industries and the rehabilitation of housing in old industrial settings. Relatively few city governments have devoted themselves consistently to using municipal instruments to direct the dividends of growth toward improving the economic situation of low-income people.

During the period when public housing construction flourished in Europe, municipalities led in providing the social wage. Recent policies of privatization and fiscal conservatism, however, indicate a major withdrawal by local governments from social welfare provision, although a few progressive governments in the United States (Clavel, 1986), some of the "red" municipalities of Italy and France, and local administrations in the north European welfare states continue to offer housing and services to low-income residents (Pickvance & Preteceille, forthcoming). In the United Kingdom withdrawal of the local state from its former redistributional role has been sharpest; Margaret Thatcher's "enterprise state" remains committed to encouraging private-sector activity without the imposition of either planning or linkage policies (Martin, 1986).

While municipalities have become more active and autonomous in their pursuit of growth policies, in the United States and United Kingdom national economic and political forces have restricted their freedom of action when they have tried to improve conditions at the bottom of the social hierarchy. U.S. firms, reacting to the high costs of doing business in cities with substantial welfare and social service budgets, fled to more hospitable locales, effectively punishing those municipalities substituting public benefits for private wages. Simultaneously, the federal government sharply reduced its subsidies for urban social programs. In Britain central government terminated the metropolitan governments, which it regarded as undercutting its policies, and severely limited the financing powers of local government, effectively preventing it from taxing its constituents to pay for higher levels of service. Thus, increases in municipal capacity in one arena have been balanced off by restrictions in another.

IDEOLOGICAL INTERPRETATIONS

Neither the systemic nor the localistic perspective on urban restructuring is necessarily connected with a particular ideological interpretation. Rather, ideology is associated with identification of the heroes and villains of the piece. The right attributes economic decline to overpaid and unproductive workers, governmental welfarism, insufficient incentives to entrepreneurship, and political intrusion into the market. While, according to this view, much of the fault lay outside municipal boundaries in national unionism and the welfare state, it also had specifically local roots:

> Current legislation to reduce local government autonomy is only the latest episode [for the Thatcherites] in a recurrent problem [with local political consciousness]. Those political processes leading to "municipal Marxism" in Sheffield or Lambeth can be replaced by "neutral" equations and civil service procedures in Marsham Street [where the Department of the Environment is located]; the Docklands Action Group can be shunted aside and "sensible" development, free of the inefficiencies of local politics, can be undertaken by the Docklands Development Corporation [Duncan & Goodwin, 1982, p. 94].

The right's prescription for regeneration therefore requires shifting the role of government from inhibitor of growth to provider of incentives. Its ideological triumphs have been enhanced by the recent introduction of market processes to the eastern socialist economies and the explicit admissions of economic failure by the Eastern Bloc leadership. Whatever the weaknesses of the logic of free markets as the basis for renewal, the right can point to the failures of communism and, in the United States and Great Britain, the economic stagnation that occurred under Democratic and Labour governments. The strength of mixed economies in northern Europe is usually underestimated or ignored in these arguments.

For leftists, urban restructuring has been produced by the greed of corporate capitalists rather than as a necessary response to the heavy hand of the state. Its outcome has been increased wealth for investors, particularly financial and real-estate speculators, and impoverishment of a growing proportion of the population. The stimulus for the process was an initial crisis of profitability caused by international competitive pressure resulting from unmanaged international trade and overproduction. Capital responded by heightening the rate of exploitation of labor. A combination of tactics was used to achieve this end, including union busting, automation, relocation of production sites, and reduction of social welfare programs that competed with the private wage (Harrison & Bluestone, 1988). At the local level capitalists worked through urban growth coalitions to establish environments favorable to cost-cutting through reduced expenditures on labor, taxes, and physical plant. Nationally and internationally they sought the locales promising the greatest return on investment, whipsawing one against the other.

According to this analysis, the effort to alleviate the situation in which most people are seeing increased insecurity and declining living conditions requires far more than a strategy for growth. Rather, it necessitates finding a formula that will limit capitalist hegemony within both the workplace and the community. No benign assumption can be sustained that, once economic growth is reestablished in a city, wage increases and a growing public fisc will follow. For instance, the recent Boeing Aircraft strike in Seattle, Washington, illustrates the way in which business seeks to exclude labor from the gains of growth. Boeing, which is the largest U.S. exporter and which currently has a huge backlog of orders, was offering a pay raise of only 10% over three years, based on the grounds of needing to compete internationally. Its union contended that when the firm was having hard times, it had accepted give-backs; now that the company was enormously profitable, the question was entirely one of the division between wages and profits, not of the need to sacrifice in order to promote competitiveness (Uchitelle, 1989c).

The fiscal effects of economic growth are similarly contested. For example, New York City business leaders, after a decade of economic growth, were pressing for across-the-board tax reductions despite dramatically growing service needs. And, in the face of serious budgetary shortfall, New York State was embarking on the third phase of a multiyear tax-reduction program, which had originally been enacted on the basis of mistaken estimates of revenue expansion. Again, the justification is competitiveness. A report sponsored by the top executives of New York City's leading firms declared:

> The increase in competitive pressures in the financial services industry has made firms more cost conscious than ever before. While cost control has always been important in choosing locations for back offices and data centers, front offices are also increasingly concerned about operating costs. . . . New York City and State impose the highest tax burden in the nation. Reducing this burden is the most important step that the city and state can take to reduce the cost of doing business in New York City, and thereby retain and promote financial services job growth in New York City [New York City Partnership, 1989, p. 23].

The triumph of progressive regimes is as important as a successful growth strategy to the well-being of citizens of declining cities. Indeed, economic stagnation may well be preferable to development if the latter is based on ruthless tax and cost reductions. The policies of progressive regimes involve public and/or worker participation in economic decisionmaking, emphasis on indigenous business development, linkage policies, housing subsidies, and a stress on neighborhood over downtown development. For cities to escape from total determination by outside forces, local entrepreneurship, planning, and distributive policies are a necessary condition. The character of the local regime determines whether, and how, these functions are carried out. But, in the old Marxist phrase, "not under conditions of their own making."

GROWTH AND EQUITY

The flaw of the leftist analysis, with which this chapter is otherwise generally sympathetic, is that it does not offer a formula for growth. So far, the left has not discovered an effective method for stimulating substantial investment in declining areas that differs significantly from the business subsidy approach of the right.

There are four conceivable sources of risk capital for economic regeneration: the private, for-profit sector; the state; employee savings and benefit funds; and the nonprofit sector. To attract private capital to territories not regarded as inherently profitable by capitalist managers, state officials feel compelled to offer incentives with all the likely negative consequences outlined above. State participation in quasi-governmental corporations has saved failing industries and is more amenable to public control of the outcomes than is state subsidy of purely private entities. (AMTRAK, the U.S. passenger railroad corporation which connects a number of old U.S. central cities and whose revival has spun off an important employment and retailing multiplier, is a good example of revitalization through the use of this kind of instrument.) Such corporations, though, when they are profitable and capitalized on a large scale, tend to behave little differently from private firms (Rueschemeyer & Evans, 1985, pp. 57-59) and, when not restricted locationally, will also seek least-cost locations. In contrast, firms run directly by the state will be less profit-oriented and, theoretically at least, susceptible to democratic control. They tend, however, to avoid risks, invest insufficiently, and avoid cost reduction measures.

Employee-owned firms offer the greatest potential for maintaining efficient operations without abandoning geographic locations that have been the site of private disinvestment. Employee takeovers, however, obviously can occur only when a firm is already in existence and must usually confront heavy debt encumbrance. Economic development corporations and economic cooperatives of various sorts have opened new enterprises and prospered in different places ranging from Chicago to Mondragon to Emilio-Romagna (Piore & Sabel, 1984). Except in Italy, however, their total contribution to economic development is tiny, and few locales possess established traditions of this sort of enterprise.

A workable, large-scale strategy based on local economic development corporations or cooperatives needs the formation of new kinds of credit institutions. Eisinger (1988, chap. 10) lists a number of innovative development banking institutions that now exist in the United States to provide loans to small businesses. Although the amount of capital so far expended is very small, Eisinger anticipates long-term cumulative benefits. As presently constituted, however, these loan funds simply do what private banking institutions do at a higher level of risk. Such funds, if constituted on more progressive principles, would issue loans containing assurances that successful firms could not be bought out, then moved or folded up. Alternatively, they could include recapture provisions such that profits from a buy-out would revert to the community.

The left needs to devise its own version of the public-private partnership. This means a reorientation away from manufacturing toward the service sector, recognition of the importance of management and entrepreneurship, and a coming-to-terms with the multinational corporation. The reality that giant multinational corporations dominate economic transactions means that the left must find ways of tapping into their economic power rather than dismissing them on moral grounds. Public-private partnerships under these conditions are inevitable; what needs to be done is insure that the public component is more controlling and shares more in the proceeds.[8]

Romanticization of the Italian machine shop cooperatively run by worthy artisans will not suffice as a model for development of old inner cities. Public sponsorship of consulting, computer, high-tech, restaurant franchise, nursing home, home health care, and similar enterprises could generate a stable, small business sector to occupy inner-city sites. If such businesses are to thrive, they will involve internal hierarchies with sufficient returns to managers as to induce competent, experienced individuals to assume these roles. They will also have to allow managers discretion in rewarding worker performance. Social equalization, if it is to occur, would come through redistribution within the tax and welfare system rather than the firm. In other words, the left will have to accept serious inequalities in the rewards to labor if it is to stimulate growth.

The task for progressive movements within declining cities is to formulate a strategy that is as creative and less destructive than the *modus operandi* of typical urban growth coalitions. Social democrats need to do what is necessary to foster incentives and reward entrepreneurship. Without a program for growth, except in cities like Santa Monica or Toronto that have to fend off private capital, the left has little chance of achieving political power. Criticisms of the depredations caused by unregulated capital or prescriptions for cooperative industry are insufficient. Most people will accept growing inequality in preference to stagnation or absolute decline in the standard of living.

To speak, however, of the tasks for progressive local forces without noting their national context is to dodge a central issue. Cities are limited in their autonomy not only by general economic forces but also by the national political system of which they form a part. Ideological, institutional, and fiscal factors constrain their ability to operate in political isolation from the rest of the nation. Within the United States and the United Kingdom, where conservative forces dominate nationally, local regimes with a different agenda must swim against the ideological current. The trickle-down model dominates the definition of economic improvement in these countries, causing other methods to be automatically suspect. Progressive local forces have difficulty maintaining a broad base of support when the national propaganda attack pictures them as loony or unrealistic. In the continental European states, where planning and social welfare maintain much greater national legitimacy, national regimes are less inclined to glorify the unshackled free market and, therefore, they give localities greater capabilities for managing development.

The extent of local entrepreneurship also depends on the amount of institutional decentralization existing within a nation's urban system. In the United States the federal system and the widespread acceptance of "home rule" have both heightened interurban competition and given cities considerable leeway to determine their own policies (Fainstein & Fainstein, 1989). In contrast, Britain's increasingly centralized system has blocked radical local councils from pursuing their own development and expenditure policies (Lawless, 1987). France, previously the most centralized of European nations, has gone through a period of decentralization that has made possible more active local efforts to foster growth.

Finally, the availability of funds to local governments and the terms under which they can use them significantly affects their capabilities. Subnational governments in the United States can tap into national financial resources through issuing tax-free industrial revenue bonds, but they have no similar source of funds for job-training and placement programs. There have been sharp reductions in direct federal support of local development and welfare programs, but localities may use their own tax revenues as they please. In contrast, in the United Kingdom, local councils, which have also seen major cutbacks in national subventions, are largely prevented from making up for the shortfall locally.

In conclusion, then, we have seen an upsurge in public-sector entrepreneurship and considerable variation in the extent to which local governments have sought to spread the benefits of growth to the whole population. We can propose programs that will increase the public benefits of growth even while encouraging private sector participation in regeneration activities. Without a broad national movement to support such programs, however, we must expect that local initiatives will be blocked by higher levels of government and by footloose capital that will play one locality against another. Entrepreneurship by urban progressive coalitions thus requires that they aim not only at stimulating local investment but also at building a national movement for growth with equity.

NOTES

1. There are two variants of the systemic perspective. The first, exemplified by Noyelle and Stanbach's (1984) study of the changing American urban hierarchy, emphasizes impersonal economic forces (increasing size of markets; changes in transport and technology; increased importance of public and nonprofit sector activities; corporate concentration) that produce economic growth and decline in particular locations. The second, most strongly presented in the work of David Harvey (1985), stresses the role of the capital-controlling class in maximizing profits through use of the "spatial fix." For Harvey, uneven development is not an unintended consequence of investment processes; rather capitalists create and use it so as to lower production costs, protect themselves from regulation, increase profitability, and produce speculative gains.

2. This kind of analysis is susceptible to the criticism that it assumes any phenomenon fulfills a necessary function and that any existing institution or activity had to be. One need not, however, engage in a totally deterministic argument to accept that certain social practices do serve the ends of dominant groups and that these practices can be institutionalized so that concerted, conscious activity is not required to perpetuate them. Because some system requirements are fulfilled does not mean that all are, and when they are, the outcome is not an automatic response to need, but can ultimately be traced back to human agency. Moreover, practices may also exist that are dysfunctional for achieving the aims of dominant social interests, and systems produce contradictions as well as functionalities.

3. A recent pair of articles in *The New York Times* (Uchitelle, 1989a, 1989b) chronicled the globalization of the Stanley Tool Company, an old New England manufacturer of screwdrivers, tape measures, and other common tools. Even though it made a seemingly low-tech, standard product, Stanley needed to be close to its foreign markets:

> The . . . [tape measure's] popularity at home raises the possibility that Stanley could increase production in New Britain [Connecticut], which turns out 200,000 tape rulers a day for the American market, and simply export the rest from here. But this approach violates principles held by Mr. Ayers [the chief executive officer] and other advocates of global manufacturing.
>
> One is that factories should be close to the customers they serve—to get inside tariff barriers, to give the impression that they are local companies, to reduce delivery time, and to "capture" manufacturing techniques not readily available back home.
>
> Another is that big factories are inefficient. The maximum for Mr. Ayers, and for many others intent on globalizing, is 500 employees, the number at the tape factory here [in France].
>
> Finally, Mr. Ayers does not like to have, as he puts it, all his eggs in one basket. "With one plant, if there is a strike or shutdown you're out of business," he said. "With several, you can switch production to another country or back and forth among countries." [Uchitelle, 1989a, p. 10].

4. The problem is more acute in American than European cities, because the jurisdictional fragmentation of U.S. metropolitan areas means that even if new functions arise in an area (e.g., Greater Cleveland or Greater Saint Louis), they may be located outside the boundary of the central city and spin off few benefits for its residents.

5. Peterson (1981) argues that growth and redistribution are necessarily antagonistic at the urban level and that the general consensus in favor of growth therefore precludes local redistributional activity, or at any rate should do so. The literature disputing his argument is by now vast, particularly concerning whether growth strategies really are supported by a consensus on values (Fainstein & Fainstein, 1986; Sanders & Stone, 1987).

6. The active agency in promoting growth within metropolitan areas in both the United States and Europe is often an intermediate level of government (states or regional authorities).

7. Quotations are from interviews carried out by the author in 1989.

8. Robert Beauregard (1989) discusses the importance of the state playing a role in requiring preferential hiring agreements for residents when it participates in development. His analysis is restricted to construction hiring, but the principle can be extended to operating firms. Another example of the public capture of benefits from major private investment is the Battery Park City project in New York. Since a public authority maintains ownership of the land, it receives an escalating rental based on profits from the structures constructed. More than $1 billion of this revenue is currently designated for low-income housing construction. This is far greater than the amounts typically allocated under linkage programs.

REFERENCES

Beauregard, R. A. (1989). Local politics and the employment relation: Construction jobs in Philadelphia. In R. A. Beauregard (Ed.). *Economic restructuring and political response* (pp. 149-180). Newbury Park, CA: Sage.

Clavel, P. (1986). *The progressive city*. New Brunswick, NJ: Rutgers University Press.

Duncan, S. S., & Goodwin, M. (1982, January). The local state: Functionalism, autonomy and class relations in Cockburn and Saunders. *Political Geography Quarterly, 1*, 77-96.

Eisinger, P. K. (1988). *The rise of the entrepreneurial state*. Madison: University of Wisconsin Press.

Fainstein, S. S., & Fainstein, N. (1989, September). The ambivalent state: Economic development policy in the U.S. federal system under the Reagan administration. *Urban Affairs Quarterly, 20*, 41-62.

Fainstein, S. S., & Fainstein, N. (1988). Technology, the new international division of labor, and location: Continuities and disjunctures. In R. A. Beauregard (Ed.). *Economic restructuring and political response* (pp. 17-40). Newbury Park, CA: Sage.

Feagin, J. R. (1988). *Free enterprise city*. New Brunswick, NJ: Rutgers University Press.

Foglesong, R. E. (1986). *Planning the capitalist city*. Princeton: Princeton University Press.

Harrison, B., & Bluestone, B. (1988). *The great U-turn*. New York: Basic Books.

Harvey, D. (1989). *The condition of postmodernism*. Oxford: Basil Blackwell.

Harvey, D. (1985). *The urbanization of capital*. Baltimore: Johns Hopkins University Press.

Hoyle, B. (1988). Development dynamics at the port-city interface. In B. S. Hoyle, D. A. Pinder, and M. S. Husain. *Revitalising the waterfront* (pp. 3-19). London: Belhaven Press.

Lawless, P. (1987). Urban development. In M. Parkinson (Ed.). *Reshaping local government* (pp. 122-137). New Brunswick, NJ: Transaction Books.

Martin, R. (1986). Thatcherism and Britain's industrial landscape. In R. Martin and B. Rowthorn. *The geography of de-industrialisation* (pp. 238-290). London: Macmillan.

Mingione, E. (forthcoming). *Fragmented societies*. Oxford: Basil Blackwell.

New York City Partnership, Financial Services Task Force. (1989). *Meeting the challenge: Maintaining and enhancing New York City as the world financial capital*. New York: New York City Partnership.

Noyelle, T. J., & Stanback, T. M., Jr. (1983). *The economic transformation of American cities*. Totowa, NJ: Rowman and Allanheld.

Parkinson, M., & Judd, D. (1988). Urban revitalisation in America and the U.K.—the politics of uneven development. In M. Parkinson, B. Foley, & D. Judd. *Regenerating the cities* (pp. 1-8). Manchester, UK: Manchester University Press.

Peterson, P. (1981). *City limits*. Chicago: University of Chicago Press.

Pickvance, C., & Preteceille, E. (forthcoming). *State and locality: A comparative perspective on state restructuring*. London: Francis Pinter.

Piore, M. J., & Sabel, C. F. (1984). *The second industrial divide*. New York: Basic Books.

Rueschemeyer, D., & Evans, P. B. (1985). The state and economic transformation: Toward an analysis of the conditions underlying effective intervention. In P. B. Evans, D. Rueschemeyer, & T. Skocpol. *Bringing the state back in* (pp. 44-77). Cambridge: Cambridge University Press.

Riley, R., & Shurmer-Smith, L. (1988). Global imperatives, local forces and waterfront redevelopment. In B. S. Hoyle, D. A. Pinder, and M. S. Husain. *Revitalising the waterfront* (pp. 38-51). London: Belhaven Press.

Sanders, H. T., & Stone, C. N. (1987, June). Developmental politics reconsidered. *Urban Affairs Quarterly, 22*, 521-539.

Savitch, H. V. (1988). *Post-industrial cities*. Princeton: Princeton University Press.

Smith, M. P. (1988). *City, state, and market*. Oxford: Basil Blackwell.
Squires, G. D. (Ed.). (1989). *Unequal partnerships*. New Brunswick, NJ: Rutgers University Press.
Sternlieb, G. (1986). *Patterns of development*. New Brunswick, NJ: Rutgers University Center for Urban Policy Research.
Uchitelle, L. (1989a, July 23). The Stanley Works goes global. *The New York Times*, sec. 3.
Uchitelle, L.(1989b, July 24). Only the bosses are American. *The New York Times*, p. D1.
Uchitelle, L. (1989c, October 12). Boeing's fight over bonuses. *The New York Times*.

Part I

Strong Leadership and Sustained Regeneration

3

Pittsburgh's "Third Way": The Nonprofit Sector as a Key to Urban Regeneration

ALBERTA M. SBRAGIA

THE LONG HISTORY OF REDEVELOPMENT

Pittsburgh has undergone such a long process of regeneration that Roman numerals and adjectives are used to distinguish particular phases in the process—"Renaissance I," "Renaissance II," and so forth. Cooperation between public and private sector elites has been so sustained that "public-private partnerships" have become a natural feature in the city's political and policy landscape.

Pittsburgh's long engagement in economic development may explain why it seems to stand apart from other American cities. In contrast to most of them, Pittsburgh has developed a public sector that is sufficiently powerful and professionalized to act as a true partner with the private sector. Neighborhoods exercise real influence on city government, unlike many cities in which downtown interests dictate what is to happen in economic development.

The redevelopment of Pittsburgh has gone on so continuously that contemporary strategies and coalitions are clearly linked to their predecessors. The public sector's strategy in the 1980s, directed toward shifting from a manufacturing base to one reliant on advanced technology, could take cooperation with the private sector largely for granted. As a result, city leaders could give considerable attention to creating the cooperation between the public and nonprofit sectors that is essential for encouraging the development of an advanced technology economic base. In many other cities,

---

AUTHOR'S NOTE: I am very grateful to Ralph Bangs, Robert Beauregard, Morton Coleman, and Edward Muller for their comments on previous versions of this argument.

cooperation with the private sector was problematic and therefore a preoccupation of their leaders.

An analysis of Pittsburgh's attempts at regeneration consequently resembles an archaeological dig. The recently completed "Renaissance II" would not have occurred as it did without "Renaissance I." Similarly, city officials' current attempts to involve some organizations in the nonprofit sector in economic development are clearly inspired by the city's Renaissance experience. Since it is the culture of both public and private elites, forged in the 1940s, that has given Pittsburgh policy-making a distinctive character, successfully inducting nonprofit-sector elites into that culture represents the city's new challenge.

The latest strategy is directed toward a transformation of the economic base, with advanced technology as the desired goal.[1] The city's two major research universities in particular play a crucial role in implementing such a strategy, for potentially, they both produce such technology and attract private-sector firms that can commercialize it. In contrast to the public-private partnership, which did not seek to enhance the economic performance of private firms, the city now uses public resources to help key nonprofit institutions enhance their role as exporters of services and importers of new advanced-technology firms.[2] The Pittsburgh Technology Center is the concrete symbol of the new and still-fragile cooperative relationship ("partnership" would be far too strong a term) between the city government and the research universities, which lies at the core of the city's current economic development strategy.

At the very least, Pittsburgh officials' explicit incorporation of the nonprofit sector into the city's economic development strategy is noteworthy. Although nationally the independent sector is an important economic actor, many analyses still assume that the American economy can be understood solely by examining the public sector on the one hand and the private (profit-making) sector on the other.[3] Urban analysts, for their part, tend to ignore the land development, employment, and export functions that universities and hospitals, for example, can play in the local economy.[4] If, over the long term, city officials do in fact succeed in fashioning cooperation among public, private, and nonprofit-sector elites in the area of economic development, Pittsburgh may well become an even more striking exception to the generalizations often made about governance in American cities.

THE ECONOMIC TRANSFORMATION OF THE CITY

In popular myth, Pittsburgh, the steel mill town, rose in the late 1980s from the debris of a collapsed steel industry with a dramatic skyline, outstanding cultural offerings, one of the best urban public education systems in the country, and extremely livable neighborhoods. However, much of the "crisis" associated with Pittsburgh's economy has more to do with the fate of the region than with the fate of the city itself. Although the economic destinies of the region and the city need to be considered together, it must be remembered that the two are not synonymous.

The central city had been deindustrializing its economy for decades at a far greater rate than its surrounding region. Consequently, by the early 1980s, Pittsburgh was not the steel mill town of popular imagery. Although the collapse of the steel industry certainly hurt the city badly, Pittsburgh possessed potential sources of new employment and therefore was able to develop new "exports." By contrast, neighboring mill towns were so dependent on steel that the industry's collapse left them reeling throughout the 1980s. They have been unable to find new exports to replace the steel that had given them prosperity for so many generations.[5]

The city has gradually lost its manufacturing base since the end of World War II. Total employment in manufacturing declined 75% from 1940 to 1980 while service employment grew 28%. In 1950, 26% of the city's jobs were in manufacturing, with 12.8% of the labor force employed in steel and metal manufacturing. By 1980 only 14.6% were in manufacturing with only 5.5% of the city's labor force employed in steel and metal manufacturing (Jezierski, 1987). By 1980 the small size of Pittsburgh's manufacturing labor force made its economy more similar to that of Boston, Atlanta, and Denver than to Cleveland or Milwaukee (U.S. Bureau of the Census, County and City Data Book, 1983).

In the period 1979-1988, the Pittsburgh region lost 100,000 manufacturing jobs, a decline of 44%. (By contrast, national employment in manufacturing decreased only 7% in those years; Giarrantani & Houston, 1989). Even more dramatically, the Pittsburgh region lost just slightly less than 88,000 jobs from 1979 to 1983, with roughly 83,000 of those in the durable goods sector. Three-quarters of the 88,000 jobs lost were related in some way to steel (Giarrantani & Houston, 1988, p. 63). By 1983, 19% of employment in the Pittsburgh region was based on manufacturing while the national figure was 21%; by 1988 only 14% of the region's employment was in manufacturing while the national figure was 19% (Giarrantani & Houston, 1988, p. 58; Sheehan, 1989, p. 4). In sum, the Pittsburgh region is no longer counted among the country's industrial regions. Its transformation was brutally swift, taking roughly eight years.

This transformation was produced primarily by the loss of manufacturing jobs rather than the growth of nonmanufacturing employment. Along similar lines, the dramatic drop in unemployment rates, which occurred after 1984, was due more to outmigration than to the growth of new jobs. Giarrantani and Houston conclude that:

TABLE 3.1
Employment in the Pittsburgh Metropolitan Region:
Years of Cyclical Peaks and Troughs for Manufacturing
(thousands of jobs)

	1966	*1972*	*1974*	*1976*	*1979*	*1987*	*1988*
Manufacturing	293	256	266	218	225	124	125
Durables	248	209	221	176	181	88	89
Nondurables	45	47	45	43	44	36	36
Nonmanufacturing	529	605	626	643	689	729	747
Total	828	861	892	861	914	853	872

SOURCE: Adapted from Giarratani & Houston (1989, p. 55).

> From 1982 to 1988 unemployment in the region declined by 67,000 workers, but 47,000 or 70 percent of this was the result of a decreasing labor force while only 20,000 or 30 percent represented new jobs. To focus only on the latter is to miss the important adjustments from outmigration and discouraged workers [Giarrantani & Houston, 1989, p. 58].

In fact, the Census Bureau has ranked the Pittsburgh region as leading the nation in population lost between 1980 and 1988.[6]

From 1987 to 1988, however, both manufacturing and nonmanufacturing jobs grew. Economic health seems to be returning. Table 3.1 traces both the decline and the recent increase in employment in the region.

Based on the 1988 increase in manufacturing jobs shown in Table 3.1, it is thought by some analysts that population losses should soon cease, and that the outlook for the region is far more optimistic than it was in the pre-1987 period. Others are more skeptical.[7] Whether the increase in jobs from 1987 on will give the region a renewed basis for economic prosperity is still unclear.

While new jobs are being created in both manufacturing and nonmanufacturing in most areas of the region, job creation is not occurring in those areas of the region where steel production had been concentrated. In the Mon Valley, where steel had reigned, the economic situation is still bleak. Between 1984 and 1988, while all other areas of Allegheny County, Pittsburgh's home county, enjoyed job growth, the Mon Valley lost 17,700 jobs (Bangs & Soltis, 1989, p. 2). Although by June 1989 the county's unemployment rate was down to 3.7%, many of the 43,600 county residents classified as unemployed, underemployed, or in low-paying jobs were in the Mon Valley.

By contrast, the city did not have a net loss of jobs nor was it left behind once the process of net job creation began. From 1980 through 1987, approximately 280,000 people worked in the city, while in 1988 the figure rose to 300,000 and is projected to be 310,000 in 1989 (Sheehan, 1989, p. 1). Growth in the city's employment base occurred primarily in 1988 and "was in many industries, including services (e.g., health, education, communications, food), manufacturing, advanced technology, and retail trade" (Bangs & Soltis, p. 7). A study carried out by researchers at the University of Pittsburgh concluded that "the county does not face the problem of many other urban areas in which the central city is declining and only suburban areas are growing. Both central city and suburban areas are growing" (Bangs & Soltis, 1989, p. 11). The study goes on to recommend that because of the rapid growth being experienced, "speeding up general job creation is unnecessary and may be harmful in a tight labor market" (Bangs & Soltis, 1989, p. 12).

The decline in the city's manufacturing employment in the 1980s has been so dramatic that the recent gain in its manufacturing employment has received special notice. It is the first yearly gain since 1967. Not surprisingly, the 1,300 new manufacturing jobs created between 1987 and 1988 have been viewed as particularly hopeful. It should be noted that since such jobs are not primarily in production, they are contributing to the "whitening" (or perhaps the "pinkening") of the city's collar. The new "manufacturing" employment in the city does not represent a return of the well-paid blue-collar job. For example, most of the 762 new jobs in the primary metals industry from 1987 to 1988 have been concentrated in the central business district. Thus, the increase did not help two of the three working-class neighborhoods for which production employment had been important in 1980.[8]

The importance of nonproduction manufacturing activity—such as management, finance, legal work, or sales—for manufacturing firms in the city of Pittsburgh is indicated by the fact that the most important manufacturing neighborhood by far is the central business district. If we assume that roughly all the jobs in the CBD are white-collar and all the jobs in neighborhoods are blue-collar, the disparity in numbers is telling. Whereas the CBD in 1988 had 16,866 manufacturing employees, the neighborhoods in the aggregate had 13,600.[9] As manufacturing jobs, both production and nonproduction, declined in Pittsburgh, they were also decentralizing. The new jobs that were created in 1987-88 were largely in small firms. The number of manufacturing firms increased from 654 to 801 from 1987 to 1988, while at the same time firms that were important before 1980 "have either reduced their operations in the city, relocated parts of or entire operations outside the city, or have gone out of business" (Urban Redevelopment Authority of Pittsburgh, 1989, p. 13).

These figures begin to indicate the startling transformation of the *kind* of economy the city of Pittsburgh now possesses. By the time manufacturing increased in 1988, the city's employment profile had become transformed. Whereas industry was the biggest employer in 1978, the universities and hospitals were in 1988. In 1978 the top seven industrial firms employed 21,108 people, whereas by 1988 that figure had declined by 47% to 11,126 persons (Sheehan, 1989, p. 1). In 1980 the central business district's manufacturing employment (all of which would have been in nonproduction-oriented manufacturing) was recorded as 25,778, whereas in 1988 that figure had plummeted to 16,866 (Urban Redevelopment Authority of Pittsburgh, 1989, p. 12).

By contrast, between 1978 and 1988 the six major educational institutions increased their employees by 26% (most of the increase was accounted for by the University of Pittsburgh and Carnegie Mellon University), while the eight most important hospitals increased employment by 22%. In 1988, then, roughly 11,000 worked for the top seven industrial firms, 37,000 for the major universities and colleges, and 23,400 in the eight major hospitals.

As the nonprofit sector has risen in economic importance, the importance of the industrial corporation has declined. Corporate takeovers and mergers have hurt Pittsburgh's standings as a city of *Fortune* 500 corporate headquarters. Gulf, for example, no longer is based in Pittsburgh because it was taken over by Chevron. In 1989 *Fortune* ranked Pittsburgh in fifth place as a *Fortune* 500 headquarters city, down from the third place it had occupied for many years. In 1989 it ranked behind New York, Chicago, Dallas, and Houston. The hemorrhage has not been as severe as those figures might indicate, however. *Forbes* in 1989 ranked the city ninth, up from the tenth place it had previously occupied.[10] The private sector as a whole remained surprisingly healthy.

Although the private sector continued to contribute significantly to the city's economy, the decline of the industrial coporation was extremely worrisome for city officials. Since industrial firms are typically the most important exporters within a city's private sector, one would expect the "export" function of the city to decline if industrial firms declined (even if the private sector as a whole remained robust). That is, fewer of the city's products would be exported outside its region, thereby slowing wealth creation.

However, while the export capacity of the city's private sector has fallen, that capacity has increased in the city's nonprofit sector. The following figures are indicative of the city's new export profile. Whereas the University of Pittsburgh's medical complex brought in only $30 million in medical research funds in 1980, by the end of 1988 that figure had risen to $91 million, much of it from the National Institutes of Health (Guo & Rouvalis, 1989, p. 6). Presbyterian University Hospital, part of the University's medical complex, alone received approximately $150 million a year in revenue from patients outside the Pittsburgh area.[11] Total sponsored research at the University of Pittsburgh rose from $58 million in 1980 to $143 million in 1988.[12]

Similarly, Carnegie Mellon University brought in $29 million in research funds in fiscal year 1979-1980, but nearly $121 million in fiscal year 1988-1989.[13] In 1985 the university won the national competition for the U.S. Department of Defense's Software Engineering Institute, a prize worth roughly $100 million of federal funds (Ahlbrandt & Weaver, 1987, p. 455). Giarrantani and Houston conclude that hospitals and universities in Pittsburgh, along with the city's banks, "were at one time solely dependent on demand generated locally by manufacturing activities and populations in the surrounding region. Now, however, they have assumed, at least in part, the character of exporting industries in the more traditional sense" (Giarrantani & Houston, 1988, p. 66).

RENAISSANCE I & II: REDEVELOPING REAL ESTATE

In the late 1970s a new mayor, Richard Caligiuri, announced that "it was time for a second Renaissance." Similarly, the ongoing $300-million redevelopment of the University of Pittsburgh's huge medical complex is termed a "Medical Renaissance." The use of the word "Renaissance" in Pittsburgh political discourse evokes memories of Renaissance I, which has become the implicit and explicit referent for any analysis about regeneration in Pittsburgh. The "public-private partnership" that characterized Renaissance I has had a particularly enduring impact and still shapes the policy-making culture of the city.[14]

Substantively, the partnership did a great deal between 1945 and the early 1960s. Edward Muller thus summarizes Renaissance I:

> Environmental reforms (smoke and flood control), physical renewal (slum clearance, parks, office buildings, and cultural amenities), and institutional restructuring (municipal service authorities for transit, parking, sanitation, regional industrial development parks, and housing associations) reversed the deterioration of downtown Pittsburgh, maintained the commitment of major corporations to the area, and kindled a spirit of optimism throughout the region [Muller, 1988, p. 39].

Renaissance I was led by a strong mayor, David Lawrence, and a strong businessman, Richard King Mellon. Lawrence dominated the city's political landscape as no politician has since then.[15] He helped establish the public-private partnership upon his election in 1945. For his part, Richard King Mellon (who died in 1970) dominated the private sector in much the same way Lawrence controlled the public sector. Mellon's commitment "to rebuild rather than abandon Pittsburgh" was the foundation stone upon which a successful partnership could be based (Lubove, 1969, p. 107). Labor, by contrast, did not try to participate in the partnership, leaving public policy decisions in Lawrence's hands. Neither member of the partnership consulted neighborhood groups because revitalization was defined as a policy for downtown.

The private sector's organizational actor in the public-private partnership was the Allegheny Conference on Community Development, organized in 1943 under Mellon's initiative. The Allegheny Conference provided the ideas, hired the professional consultants who drew up detailed redevelopment plans, and in general led Renaissance I. The public sector reacted to the plans, usually with acceptance (Lubove, 1969, p. 107). From the record, however, it does not appear that the public sector had alternative ideas as to what to do in Renaissance I. Even the slum clearance program, which dislocated thousands of mainly poor blacks, was approved by the county's first black judge. Although the private sector dominated, it was at least due as much to a lack of perceived options as it was to the power it mobilized.

Lawrence worked closely with the Allegheny Conference. In fact, when he became governor in 1958 he appointed Park Martin, the Conference's executive director, to his cabinet. He also institutionalized the public-private partnership by establishing the Urban Redevelopment Authority and appointing himself as chairman (as position he held until his death in 1966) and three influential Republican businessmen to the five-member board. The three included Mellon's key adviser. Morton Coleman concluded that "The five-member Authority became the public instrument used to implement the Lawrence-Mellon alliance. It also provided Republican economic interests with access and critical public power in decisions in the city" (Coleman, 1988, p. 135). Conversely, the corporate leadership abandoned the city's Republican Party, which had controlled the city before the New Deal.

In 1982 the city housing department was shifted to the URA so that it is now charged with all public-sector activity in the areas of both housing and economic development. It currently dominates much of the city's policy-making landscape. Lawrence's legacy is still felt: John Robin, the current chairman of the URA board, had been Lawrence's executive secretary and was appointed by Lawrence as the URA's first executive director.

When Lawrence was elected Governor of Pennsylvania, he picked Joseph Barr to succeed him as mayor. From 1959 to 1969 Mayor Barr continued the process of redevelopment, while also trying to meet the radical and social challenges of the 1960s.[16] Citizen opposition to the slum clearance that marked urban renewal during the late 1950s and 1960s and to the high priority given downtown became very organized and very vocal. By contrast, the Allegheny Conference did not respond to the urban crisis and focused exclusively on redevelopment issues until 1968.

In that year the role of the "technocrats" in the process of redevelopment and in the shaping of the partnership became crystallized. The staff, as it were, came front and center. Robert Pease was appointed as the new executive director of the Allegheny Conference—a job which he took by leaving his position as executive director of the Urban Redevelopment Authority. With his appointment, the Conference became far more active in responding to social problems. Pease, as of 1989, is still the Conference's executive director and, along with John Robin, forms a crucial part of the "institutional memory of redevelopment" in Pittsburgh.[17] In fact, he is playing a pivotal role in fashioning cooperation among the public and nonprofit sectors, an irony indeed.

Neighborhood activists found a candidate in Peter Flaherty, a loyal member of the party and Barr's heir apparent, who broke with the machine and ran for mayor without party endorsement. He stressed that he owed no allegiances—not to labor, business, or the party—and promised that neighborhoods rather than downtown would be given priority.

His election in 1969 introduced a period in which the administrative and financial structures of government were given priority, redevelopment was shelved, and the partnership lay dormant. He cut back on patronage by dramatically eliminating city jobs and shifted costly functions to the county (Clark & Ferguson, 1983, p. 194; Stewman & Tarr, 1982, p. 91). He stressed efficiency and savings in government, and did not view government either as providing social welfare or as facilitating redevelopment (Clark & Ferguson, 1983, pp. 198-203).

Flaherty *was* concerned with neighborhoods (Stewman & Tarr, 1982, p. 93). His most enduring legacy has to do with the restructuring of the city planning department carried out under his administration. The department was forced to listen to neighborhood groups (Lurcott & Downing, 1987, pp. 460-461; see also Sbragia, 1989). Under Flaherty, neighborhood groups became an important part of the city's political and policy equation. Their incorporation into the city's policy-making system laid the basis for the strikingly consensual nature of redevelopment in Pittsburgh in the 1980s.

The changed role of neighborhood groups became clear in "Renaissance II," as the period of redevelopment in Caliguiri's administration was known. Caliguiri became mayor when Flaherty joined the Carter Administration in Washington in 1977, and, although more redevelopment of the downtown area was one of his top priorities, so was neighborhood revitalization. The traditional partnership between the Allegheny Conference and the mayor's office was revived to redevelop the downtown area, but the mayor also allocated a great deal of time and money to neighborhood concerns. As the downtown became transformed by new and architecturally striking skyscrapers as well as cultural facilities, the neighborhoods were also receiving attention (Ahlbrandt & Weaver, 1987, pp. 120-134). By 1987 Caliguiri, in his own words, worked "a lot closer with . . . community organizations than . . . with the business organizations."[18]

It is important to note that, although the implementation of Renaissance II coincided with the collapse of the steel industry in the region, Renaissance II was conceived and begun *before* the advent of economic crisis. Renaissance II was designed to redevelop real estate rather than to restructure the city's economy. The gleaming skyline—representing roughly a $750-million investment—which Renaissance II produced, certainly helped the city cope with the economic problems of the 1980s. But the impact of Renaissance II on economic restructuring was coincidental.

The latest strategy for economic development moves away from the models of regeneration represented by Renaissance I and II.[19] This current strategy, which is still evolving and is more of a direction than a strategy as such, does not focus on downtown, views real estate development as facilitating strategic economic activity rather than being an end in itself, and relies on nonprofits and variants of public (i.e., at least "nonprivate") monies rather than simply on private firms and private-sector investment. Rather than transforming land use, it hopes to restructure both the economic base and the social structure of the city. The strategy, not surprisingly, is characterized by the consensual style of policy-making rooted in Renaissance II.

The lack of challenge to such a strategy is striking. The city's economic development policy is not easily compatible with the needs of those residents unequipped to function in an economy based on advanced technology. Those without education or skills will not fare well in the "new" Pittsburgh—and yet they have not mobilized. The black community in particular has been silent, even though the current strategy raises grave long-term issues for a community already suffering from serious unemployment and low labor force participation.[20]

RENAISSANCE III: NONPROFITS AS ENGINES OF GROWTH

The city needed more than its traditional public-private partnership to address the problems of the city's economic base. The partnership, as traditionally constituted, had focused on real estate development and could not address economic restructuring, especially since many private-sector members of the partnership were undergoing restructuring themselves. Furthermore, many local officials felt that, given the interregional competition faced by manufacturing in the Pittsburgh region, the disappearance of production jobs was inevitable. Therefore, attention should be focused elsewhere.[21] Although the mayor became active in efforts to use public power to restore some steel-making capacity in the region, it became accepted that, even in the best case, relatively few production jobs could be created in the city itself.[22]

In the early 1980s, just as companies were facing unfamiliar fierce competition, being taken over, or going into bankruptcy, both the University of Pittsburgh and Carnegie-Mellon University, for their own reasons, began to expand and make a concerted effort to become major recipients of research dollars. In turn, the universities and their spinoff organizations—hospitals and advanced-technology firms—began to be viewed both as major economic forces within the city and as being among the key strategic actors in economic regeneration.[23]

By 1986 the change in the city's economic base became clear in rather dramatic ways. In July of that year, the steel company that had been the city's largest single employer until the mid-1970s declared bankruptcy—and in October of that same year city, county, and university officials announced that the Pittsburgh Technology Center would be built on the site of a closed steel mill. The two universities would be the main tenants. The University of Pittsburgh would build a Center for Biotechnology and Bioengineering while Carnegie-Mellon University would construct the Carnegie-Mellon Research Institute, housing the Center for Advanced Manufacturing and Software Engineering. The University would also codevelop a second building designed to house the research activities of private high-tech companies.

The Pittsburgh Technology Center crystallizes the city's attempt to move toward an economic base characterized by advanced technology and anchored to the city's two research universities. Mayor Caliguiri characterized it as "the most important economic development project undertaken in Pittsburgh in the past 40 years."[24]

The Technology Center formed part of a larger economic development package that was submitted to the Commonwealth of Pennsylvania's state legislature for funding. The package, entitled *Strategy 21: Pittsburgh/Allegheny Economic Development Strategy to Begin the 21st Century* and known simply as "Strategy 21," both crystallized and deepened the growing interdependence between the public and nonprofit sectors.

The package, which eventually obtained nearly $70 million in state funds for those projects destined for the city, was put together very quickly. Under the guidance of Robert Pease of the Allegheny Conference, it was first discussed in March 1985, at the prodding of state legislators from Western Pennsylvania, and the entire package was made final on April 26, 1985. Pease (who had been asked, by the most powerful county commissioner, to coordinate the process) included, in the agenda-setting stage, representatives from city and county agencies, the University of Pittsburgh and Carnegie-Mellon University as well as city and county lobbyists and the mayor's executive secretary. These staff members put together a package and presented it at a "summit meeting" attended by the mayor of the City of Pittsburgh, the Allegheny County Commissioners, a representative of Carnegie-Mellon's president, and the president of the University of Pittsburgh.

There, the universities' proposals for a biotechnology center and a robotics center were added to the package proposed by the staff, which included the Pittsburgh Technology Center (which had been under discussion since 1983), funding for the two universities' advanced technology work separate from the biotechnology and robotics centers, a new Carnegie Science Center to be constructed in the city, and a major expansion of the airport. The airport expansion was the most expensive proposal ($100 million) and was viewed as the top priority for the region, while the Technology Center and the university centers as well as the Carnegie Science Center were the most important for the city. The city's future development was clearly intertwined with the agendas of the universities and the science center—all nonprofits.

Both the staff and the principals agreed to the final package in a process that achieved, in Coleman's words, "consensus . . . through polite bargaining."[25] Since the Strategy 21 document did not indicate priorities, each participant was able to include those projects of top priority to him/her and accept the inclusion of the top priorities of the other participants.

The process for formulating the final package involved bargaining rather than planning in the sense that planners would define it. Nonetheless, such bargaining forced actors who are often suspicious of one another to listen and respond to each other. The county and the city had to cooperate with each other as did the two universities. Government-nonprofit cooperation was explicitly addressed. All the participants had to drop certain projects, and, for the first time, discussion took place about the shape that public investment in the region and the city should take.

CAN GOVERNMENT AND NONPROFITS COOPERATE?

The relationship evolving between the city and the universities in the process of actually establishing the Technology Center can provide the basis for future cooperation. Nonetheless, the future shape of the government-nonprofit relationship remains unclear. Although city officials think the universities and hospitals are key actors in economic development, such recognition is not sufficient to institutionalize the kind of cooperative arrangement that gave birth to the Technology Center. Whether the Technology Center will be followed by other cooperative ventures depends both on city-university relations and on inter-university relations.

Relations between governments on the one hand and nonprofits on the other are often tense, with conflicts erupting over the latter's tax-exempt status and requests for zoning changes. Pittsburgh is no exception in this respect (Flaherty et al., 1990, p. A1). The city is currently exploring ways of obtaining the equivalent of tax revenue from both universities and hospitals. The latter are generally extremely resistant to opening the door to any form of taxation.

Furthermore, no institution created serves the function that the Board of the Urban Redevelopment Authority served for the Lawrence-Mellon partnership. There is no institutionalized forum that allows bargains to be made between the government and the nonprofit sector. Similarly, no forum exists that would institutionalize regular dialogue.

Surprisingly, however, achieving cooperation *among* nonprofit institutions themselves may be the most difficult aspect of creating a "government-nonprofit partnership." Two structural problems make it more difficult to achieve such cooperation than to achieve its equivalent among private firms acting through the Allegheny Conference.

Nonprofits do not have the equivalent of an Allegheny Conference to help mediate conflicts among themselves or to mediate between them and the public sector. Cooperation between the mayor's office or the Board of Education, on the one hand, and the Allegheny Conference, on the other, is a normal part of policy-making in Pittsburgh and does not usually involve any specific private firm, unless Robert Pease decides the firm should be involved. By contrast, the universities and hospitals deal individually with all other actors. Coordination between them is difficult even if the will exists. When it does not, the problem becomes severe.

The competitiveness of nonprofits vis-à-vis one another also presents a problem for an eventual partnership. The Allegheny Conference mobilized the private sector for real estate redevelopment, a goal that was incidental to the economic survival of the firms allowed to participate in the Conference. The Conference did not interfere with firms' investment strategies or other business decisions, and those businessmen, such as developers, who had a direct interest in real estate development were not allowed to join. Cooperation with one another and with the public sector through the Conference did not affect firms' competitive position within their markets.

It is unclear whether the same is true for nonprofits such as universities and hospitals. Hospitals in Pittsburgh compete fiercely for patients and research dollars. Similarly, the two universities are competitors across a wide range of areas. At a more personal level, the two universities' presidents are, at best, fiercely competitive with one another, and thus cooperation between the two institutions is always problematic. Cooperation will not always necessarily advance the interests of either individual hospitals or universities. Perhaps even more important, the perception of competition is so great that even when cooperation will not affect business operations adversely, such cooperation may be viewed with suspicion.

Pittsburgh therefore enters the 1990s facing the challenge of building new institutions suitable for a new economy in which the nonprofit sector is crucial. Institutions are needed to increase the capacity for cooperation among those nonprofits identified as pivotal for economic development; further, there is a need for institutions that can bring government and the nonprofit sector into more regular contact and negotiations. Organizationally, it will be a more difficult task than was faced by Lawrence and Mellon, for the private sector of the 1940s was far more hierarchical than is the nonprofit sector of today. The economic constraints limiting cooperation among nonprofits may be different from those that operated on private firms during the heyday of the public-private partnership.

At the very least, city officials have moved beyond an exclusive focus on supporting and bargaining with the private sector. As public-private partnerships become ever more fashionable elsewhere, Pittsburgh has already recognized that certain nonprofit organizations are economic engines, and thus should receive public-sector attention. The nonprofit sector, in brief, can be as strategic to economic development policy as is the private sector.

NOTES

1. The goal of an economic base founded on advanced technology is regionally shared, and numerous organizations are working toward this goal. Two of the most visible are the Western Pennsylvania Advanced Technology Center of the Ben Franklin Partnership and the Pittsburgh High Technology Council. For an analysis of the advanced technology sector, see DeAngelis (1989, pp. 47-48).

2. The nonprofit sector includes a wide variety of organizations. Universities, hospitals, museums, science centers, social service providers, and organizations created to develop land all belong to the independent sector. In this essay, I am concerned with institutions such as hospitals and universities rather than with, for example, community development corporations. While the latter are potentially important in stimulating neighborhood economies, they are not exporters of services and therefore do not contribute to the aggregate creation of wealth in the way organizations that export products or services do.

3. For exceptions, see Ginzberg, Heistand, and Reubens (1965), Weisbord (1977), and Salamon (1987).

4. For an exception, see Benson (1985).

5. Morton Coleman, writing in mid-1987, concluded that:

> The dominant economic/political issue of the region has shifted from the renewal of downtown Pittsburgh with its strong mayor, highly sophisticated political bureaucracy, and tradition of public/private partnership, to the issue of serious economic decline in the industrial towns along the rivers of the region and their fragmented and fiscally strained governmental systems [Coleman, 1988].

6. However, the population decline reached its peak in 1985, when 34,600 people are estimated to have left. In 1988, by contrast, only 7,900 left (see Stoffer, 1989, p. 1).

7. For two contrasting analyses, see Bangs (1989, pp. 1-16) and Giarrantani and Houston (1989).

8. In Lawrenceville, the 685 workers in primary metals in 1989 (most of whom would have been in production) were down to 92 in 1988; the figures for the Southside were 285 and 34 (Urban Redevelopment Authority of Pittsburgh, 1989, p. 7).

9. After the CBD, the next most important manufacturing neighborhoods were, in order of importance, the lower Northside with 5,743 employees, Lawrenceville with 2,383 employees, East Liberty with 1,906, Hazelwood with 1,302 (nearly all in the LTV coke works), Oakland with 1,264, and the Southside with 1,004 (Urban Redevelopment Authority of Pittsburgh, 1989, p. 12).

10. *Fortune,* which lists publicly traded industrial companies ranked by sales, found Pittsburgh to be home to 12 *Fortune* 500 companies. The *Forbes* list, by contrast, is made up of all companies that are publicly owned. Thus, banks, utilities, and other types of service businesses with sales, profits, assets, or stock value in the top 500 make the *Forbes* list. Pittsburgh was home to the corporate headquarters of 14 such companies, while Minneapolis and San Francisco have 13 each and Atlanta has 15. The contrast in direction between the *Fortune* and *Forbes* rankings confirms the notion that industrial firms are becoming less important within the private sector in Pittsburgh as that sector becomes more diversified (Gallagher, 1989, p. 1).

11. Data received from the News Bureau, Schools of the Health Sciences, University of Pittsburgh.

12. Data received from the Office of Research, University of Pittsburgh, October 10, 1989.

13. These figures were given by Carnegie-Mellon University Office of Public Relations, October 9, 1989.

14. For example, a public-private service consortium made up of the city government, nine corporate investors, and four individual investors bought the Pittsburgh Pirates to ensure that they would not leave Pittsburgh! Even cooperation among hospitals and units within the University medical complex is named the "Partnership for Medical Renaissance."

15. Lawrence was both the Pennsylvania Democratic state chairman and head of the Pittsburgh political organization or "machine", had been Pennsylvania's Secretary of State, was elected mayor of Pittsburgh four times, and then went on to become Governor of Pennsylvania.

16. For an excellent analysis of the tensions faced by the Barr Administration, see Coleman (1983).

17. The Pittsburgh experience points to the importance of staff relationships. Professionals fairly quickly came to play pivotal roles on both the public- and private-sector sides. Consequently, the culture that has emerged after decades of public-private interaction is one in which staff professionals play key roles. They do much of the bargaining and negotiation. In turn, they have helped create a policy environment that is quite different from the boisterous and raucous politics of, say, Chicago. The "politics of consensus" describes Pittsburgh politics more accurately than it does that of many other Eastern and Midwestern cities.

18. Piechowik and Perlmutter (1987, p. A12). In its 1986 Annual Report (1986, p. 2), the URA leadership stated that "the public/private partnership that the URA helped forge in Pittsburgh 40 years ago has evolved into what today might more accurately be called a 'public/private/neighborhood' partnership."

19. For a discussion of the projects of Renaissance II that received public monies administered by the Urban Redevelopment Authority, see the Urban Redevelopment Authority 1988 Annual Report (1988, p. 2).

20. For a discussion of the black community's role (or lack thereof) in economic development policy-making, see Sbragia (1989, pp. 103-120).

21. This attitude was also strong at the state level. Susan Hansen argues that:

> Harrisburg . . . seems to have accepted the viewpoint of many economists who argue that a large-scale, fixed technology industry like steel is ill-suited for competition under current circumstances, and that the state should therefore not expend resources on its behalf [Hansen, 1988, p. 107].

Hansen provides an excellent analysis of why *state* government did so little to help the steel industry and its unemployed workers. Among other reasons, she points out that the state now has such a diversified economy that "neither the steel industry nor its workers enjoy the clout in Harrisburg that they once had" [Hansen, 1988, p. 107].

For a discussion of public policy responses (or lack thereof) to the decline in the region, see, in addition to Hansen, Ahlbrandt & Weaver (1987, pp. 449-458), and Coleman (1988, pp. 123-158).

22. For a discussion of the mayor's efforts on behalf of recreating a steel-making capacity in the region, see Sbragia (1989).

23. Although the data on spinoffs are very poor, it is clear that both universities do have such spinoffs. The city's hospitals are expanding, it is felt, because of the rise of the University of Pittsburgh's reputation as a center for medical research. Carnegie-Mellon University's President claims that more than 50 firms have been created because of some link to research at that university. Recently, Pittsburgh was chosen as the headquarters for the Eastern operations of NeXT Inc., a computer firm established by Steven Jobs, the former chairman of Apple Computer. According to a company executive, the company's "collaborative relationship" with

CMU was important in choosing Pittsburgh over Boston (see Gannon, 1989, p. 9; Shehan, 1989, p. 1).

The recognition of the universities as important for economic regeneration was explicitly stated by the Urban Redevelopment Authority in its 1986 Annual Report (1986, p. 7).

24. Quoted in the Urban Redevelopment Authority of Pittsburgh, 1986 Annual Report (1986, p. 7).

25. My whole discussion of the Strategy 21 process is drawn from Coleman's analysis (1988).

REFERENCES

Ahlbrandt, R. S., Jr., & Weaver, C. (1987, Autumn). Public-private institutions and advanced technology development in Southwestern Pennsylvania. *Journal of the American Planning Association*, 449-458.

Bangs, R. L. (1989, September). Recent regional economic and demographic trends. In R. L. Bangs & V. P. Singh (Eds.). *The state of the region: Economic, demographic and social trends in Southwestern Pennsylvania* (pp. 1-16). Pittsburgh: University Center for Social and Urban Research, University of Pittsburgh.

Bangs, R. L., & Soltis, T. (1989, June). *The job growth centers of Allegheny County: Interim report for the project: Linking the unemployment to growth centers in Allegheny County.* Pittsburgh: University Center for Social and Urban Research, University of Pittsburgh.

Benson, V. O. (1985, July). The rise of the independent sector in urban land development. *Growth and change: A journal of public, urban, and regional policy, 16*, 25-39.

Clark, T., & Ferguson, L. C. (1983). *City money: Political processes, fiscal strain, and retrenchment.* New York: Columbia University Press.

Coleman, M. (1983). *Interest intermediation and local urban development.* Doctoral dissertation, Department of Political Science, University of Pittsburgh.

Coleman, M. (1988). Public/private cooperative response to regional structural change in the Pittsburgh region. In J. J. Hesse (Ed.) *Regional structural change and industrial policy in international perspective: United States, Great Britain, France, Federal Republic of Germany* (pp. 123-158). Baden-Baden: Nomos Verlag Publishers.

DeAngelis, J. (1989, September). Advanced technology businesses in Southwestern Pennsylvania. In R. L. Bangs & V. P. Singh (Eds.). *The state of the region: Economic, demographic and social trends in Southwestern Pennsylvania* (pp. 47-68). Pittsburgh: University Center for Social and Urban Research, University of Pittsburgh.

Flaherty, M. P., Twedt, S., Fraser, J., & Buell, T. (1990, February 4). Who pays? Health, wealth focus on hospital tax-break debates. *Pittsburgh Press*, p. A1.

Gallagher, J. (1989, April 10). On Forbes' new list, city takes a step up. *Pittsburgh Post-Gazette*, p. 1.

Gannon, J. (1989, August 22). NeXT to locate eastern office here. *Pittsburgh Post-Gazette*, p. 9.

Giarrantani, F., & Houston, D. B. (1988). Economic change in the Pittsburgh region. In J. J. Hesse (Ed.). *Regional structural change and industrial policy in international perspective: United States, Great Britain, France, Federal Republic of Germany* (pp. 49-88). Baden-Baden: Nomos Verlag Publishers.

Giarrantani, F., & Houston, D. B. (1989, December). Structural change and economic policy in a declining metropolitan region: Implications of the Pittsburgh experience. *Urban Studies 26*.

Ginzberg, E., Heistand, D. & Rubens, B. (1965). *The pluralistic economy.* New York: McGraw-Hill.

Guo, D., & Rouvalis, C. (1988, October 4). Detre transforms Pitt medical complex. *Pittsburgh Post-Gazette*, p. 6.

Hansen, S. B. (1988). State governments and industrial policy in the United States: The case of Pennsylvania. In J. J. Hesse (Ed.). *Regional structural change and industrial policy in international perspective: United States, Great Britain, France, Federal Republic of Germany* (pp. 89-121). Baden-Baden: Nomos Verlag Publishers.

Jezierski, L. (1987). *Political limits to development in two declining cities: Cleveland and Pittsburgh.* Paper presented at the annual meeting of the Southwestern Political Science Association.

Lubove, R. (1969). *Twentieth century Pittsburgh: Government, business, and environmental change.* New York: Wiley and Sons.

Lurcott, R. H., & Downing, J. A. (1987, Autumn). A public-private support system for community-based organizations in Pittsburgh. *Journal of the American Planning Association, 53*, 459-468.

Muller, E. K. (1988). Historical aspects of regional structural change in the Pittsburgh region. In J. J. Hesse (Ed.). *Regional structural change and industrial policy in international perspective: United States, Great Britain, France, Federal Republic of Germany* (pp. 17-48). Baden-Baden: Nomos Verlag Publishers.

Piechowiak, C., & Perlmutter, E. M. (1987, November 15). Region's fast change challenges old-boy network. *Pittsburgh Press*, p. A12.

Salamon, L. M. (1987). Partners in public service: The scope and theory of government-nonprofit relations. In W. Powell (Ed.). *The nonprofit sector: A research handbook* (pp. 99-117). New Haven: Yale University Press.

Salamon, L. M. (1987, January-June). Of market failure, voluntary failure, and third-party government: Toward a theory of government-nonprofit relations in the modern welfare state. *Journal of Voluntary Action Res., 16* (1 & 2), 29-49.

Sbragia, A. (1989). The Pittsburgh model of economic development: Partnership, responsiveness, and indifference. In G. Squires (Ed.). *Unequal partners* (pp. 103-120). New Brunswick, NJ: Rutgers University Press.

Sheehan, A. (1989, October 2). An industrial city no longer. *Pittsburgh Post-Gazette*, p. 1.

Stewman, S., & Tarr, J. A. (1982). Four decades of public-private partnerships in Pittsburgh. In R. S. Fosler & R. A. Berger (Eds.). *Public-private Partnerships in American cities: Seven case studies.* Lexington, MA: D.C. Heath and Company.

Stoffer, H. (1989, September 8). Population drop leads nation. *Pittsburgh Post-Gazette*, p. 1.

Urban Redevelopment Authority of Pittsburgh, Economic Development Department. (1989, August). *Manufacturing trends in Pittsburgh: 1980 through 1988.*

U.S. Bureau of the Census, Department of Commerce. (1983). *County and city data book.* Washington, DC: U.S. Government Printing Office.

4

Economic Regeneration in Rennes
Local Social Dynamics and State Support

PATRICK LE GALÈS

THE COMPONENTS OF RENNES' REGENERATION

Rennes belongs to a group of dynamic of French cities which had the local social resources and the local culture to take advantage of the major decentralization reforms that took place in France in the 1980s. Before the 1789 Revolution, it was a provincial capital that expanded around its parliament and university, but remained untouched by the Industrial Revolution. In contrast with Marseilles, which is experiencing an urban crisis, Rennes has recently experienced remarkable regeneration based on three factors: (1) extensive local cooperation, (2) an innovative regeneration strategy that brings together new technologies, urban planning, culture, and higher education, and (3) central state support.

POSTWAR GROWTH IN A REGIONAL CAPITAL

Rennes is the regional capital of Brittany. It stands near the border with Normandy, about 150 miles east of Brest, the western extreme of Brittany, and 230 miles west of Paris. It lies at the confluence of two small rivers, which gave the name to the *department "Ile et Vilaine,"* of which Rennes is also the administrative capital. As are most French urban areas, Rennes is divided into many independent *communes*. In this chapter, we refer to the Rennes District, which includes 26 *communes,* roughly equivalent to the urban area; however, either the *bassin d'emploi* ("travel-to-work area") or Rennes City Council, the main municipality at the center, is sometimes used.

Rennes entered the twentieth century hardly touched by the Industrial Revolution, remaining mainly an administrative center dominated by the church, the army, the university, the railways, and some isolated industries such as a brewery, a printing house, and a growing building industry. Rennes' industrial takeoff occurred after 1945, when it enjoyed almost the highest growth rate among French cities. Until the 1930s half the French population still lived in rural areas. Most urbanization took place between 1945 and 1975, when France reached a level of urbanization comparable to that of other European countries. The origins of Rennes' takeoff were twofold. First, and most important, Brittany, and Rennes in particular, were among the main targets of the state's regional policy implemented after the war. Second, traditional industries were also developed during this period.

The turning point occurred for Rennes when the Citroen car company located two plants in the area. The first was built in 1953 on land prepared by the Chamber of Commerce and Industries. The second was developed on the edge of the city seven years later. All together, 14,000 people work there, a testimony to the *deconcentration* movement initiated by the state in industries and services (Laborie, Langumier, & De Roo, 1986). Because of Rennes' rural background and the absence of a union tradition, several corporations soon set up plants within the area, including ITT (telephones), SGS ATES (semiconductors), SPLI (textiles), and Antar (oil).

Within the regional policy plans, Brittany was designated by state planners as a *vocation electronique,* justifying more deconcentration in the field. As a result, several public centers of research and *Grandes Ecoles* were decentralized to Rennes. A whole series of small companies was established in the early 1970s in the electronics sector, attracted by these research centers. Local building industries also enjoyed considerable development, supported by the steady growth of Rennes. The Rennes District population grew from under 150,000 after the war to nearly 300,000 in the mid-1970s. The agricultural modernization of Brittany also generated prosperous agricultural cooperatives. Dynamic local entrepreneurs were found in electronics and the building trades. Despite these developments and the widening of its industrial base, the city remained dominated by the tertiary sector. Jobs in the industrial sector represented about one-third of local employment in the district—about 30,000 jobs—half of them provided by Citroen.

As might be expected from a city with so little industrial tradition, private-sector elites have not played a major role in Rennes' development. Rather, it is marked by two sets of actors—banks and state agencies, and the local intellectual Christian Democrat elites. The first characteristic of urban elites in Rennes is the influence of the progressive Christian Democrat and regionalist culture, and the integration and dynamism of this network.

Brittany has been a stronghold of French Catholicism, and Rennes' traditions have been deeply influenced by the Catholic culture, especially by the humanist, Christian Democrat part of it. The local newspaper, *Ouest-France,* was founded by a priest at the beginning of the century, and the first and still most important union has been the Confederation Francaise des Travailleurs Chretiens (CFTC), later the CFDT. Rennes' leaders have belonged to this tradition of progressive Christian Democrats, including youth and peasant movements, union, and the political party which disappeared in the 1960s, the Movement des Republicains Populaires (MRP) (Teitgen, 1988).

Between 1953 and 1977 Rennes City Council was run by a Christian Democrat *grand maire,* Y. Freville, a professor at the university, and a local *grand notable* who held numerous offices. He was *depute,* then *senateur,* chairman of the District and chairman of the *departement.* Freville's election marked a turning point because he was elected as an ally to the Socialists against the local bourgeoisie—mainly shopkeepers and small entrepreneurs whose main concerns were to avoid taxes. His program had four main elements: (1) a powerful land use plan to control the expansion of the city, (2) the development of education and higher education, (4) the expansion of the infrastructure, and (4) the establishment of Rennes as an international intellectual and cultural center. In 1953 his primary goal was to attract high-technology industries by developing the university and research within the city.

This was an innovative strategy for its time and is explained by a variety of factors. For example, the Catholic influence emphasized urban planning to avoid land speculation. The regional influence encouraged efforts to bring international prestige to Rennes and Brittany, in contrast to Paris. The humanist Christian Democrat influence argued that universal education was the key to the development of a harmonious city. The academic influence contributed the view that research and the university were the key to future scientific and economic growth. Adopting these principles, Rennes became not only one of the symbols of modernist urban planning in France but also one of the most innovative councils in terms of local involvement in the economic and cultural development of the city. Despite the divisions among the *communes* of the area, Freville managed to create a powerful intercommunal body, the District, with a planning agency called AUDIAR (*Agence d'Urbanisme et d'Amenagement Intercommunal de d'Agglomeration Rennaise*).

The reign of Freville was an example of municipal presidentialism in France. His broad ideas had the support of virtually all local actors. The university, the trade unions, the local newspaper, the church, young entrepreneurs, the Chamber of Commerce, the voluntary sector, and the local officials all shared this progressive Christian Democrat and regionalist culture. This elite consensus on the main goals of economic and social development has been the distinctive feature of Rennes' political life.

The second characteristic of Rennes' urban elites concerns the role of the central state. Indeed, Rennes City Council modernist strategy could only succeed because it was matched by state policies at the same time. Gaullist elites were worried by the growth of an influential regional movement in Brittany and convinced the region had to be developed. Finally, on the political front, MRP ministers were in nearly all the national governments until 1964. In his diaries, Freville explains that he could secure government programs in housing and education, for example, because his old friends from the MRP were sympathetic to his demands.

The Gaullist regime, which started in 1958, was marked by increased state intervention. Civil servants were given powers to organize the modernization of the country, sometimes against the will of the more conservative local elites. Freville's ambitions for his city matched this strategy. He was also able to secure a close network of high-powered civil servants who contributed support for Rennes' projects (Freville, 1977). In other words, the city's most influential economic actors were civil servants, both in Paris and in various local agencies, who were implementing the national modernist strategy in Rennes.

For instance, the *Societe de Developpement Regional,* a public financial institution, was created in Rennes in 1956-1957 to finance long-term loans, with cheap interest rates, and investment for local and regional firms. The role of the *Caisse des Depots et Consignations* (CDC) was also crucial. It was the compulsory banker for local authorities. To develop a housing program, a local authority needed a loan from the CDC. In fact, it would try to get some part of a national housing program with special pubic grants, in conjunction with the Prefect. The CDC established its regional subsidiary in Rennes—the *Societe d'Economie Mixte d'Amenagement de la Bretagne* (SEMAEB), funded jointly by the CDC and local authorities. A third example of state intervention is provided by the role played by the network of *Ouest Atlantique*. This is both a regional branch of a policy agency, DATAR (*Delegation a l'Amenagement du Territoire et a l'Action Regionale,*) and a development agency supported by regional Chambers of Commerce and Industry and some private firms. Although located in Nantes, 60 miles south of Rennes, it played a major role in attracting business to the west of France, in keeping with the priorities of the state regional policy. Finally, regional civil servants, the Prefect, and his counterpart for education often supported local projects for urban planning: housing programs, establishing a second university, attracting new activities.

By contrast, Rennes' limited industrialization has led to a relatively small working-class population that hardly developed class consciousness, except in the railways. It led to limited influence for the Communist trade unions on the one hand, and the local bourgeoisie on the other. The Chamber of Commerce has been relatively weak, although it is a compulsory public body in France, which mainly brings together shopkeepers. It was only in the 1960s that local industrialists became politically active. To some extent, this limited social conflicts and social division in the city. Since the war, the dominant social group has always been the university lecturers and professors, demonstrated by the fact that all the mayors and political leaders in the main parties have come from the university.

Thus local political elites and regional civil servants were the driving forces behind Rennes' economic development. The local modernist strategy and its innovative leadership matched the state priorities of regional development and urban modernization. By contrast, the private sector played a relatively minor role, with the possible exception of the regional *Banque de Bretagne*, a national developer that worked in cooperation with the Ministry of Planning to implement a national policy in Rennes. There was no strong private-sector leadership, and besides, they opposed Freville's strategy. Until the end of his career, local shopkeepers never accepted the modernization of the city and, for example, blamed him for the development of shopping centers. Local industrialists, who also originally opposed Freville, were too weak or independent to play a dynamic role and never worked closely with the city.

THE SEARCH FOR SOLUTIONS

The crisis in Rennes in the 1970s did not take the form of a collapse similar to those experienced in American and British industrial cities. There was no port economy to collapse, and Citroen's selection of Rennes as a major center prevented there being an economic disaster in the city. Rather, its decline was marked by the slow erosion of its already weak economic base. However, the decline of the industrial sector within an area with such a fragile industrial base—it reached a significant size only in the 1970s—marked the limits of the city's development. Indeed, both the stagnation of manufacturing employment in the district and a decline in the city center were hidden by the continual increase in jobs in the tertiary sector. But the sharp decline of traditional industrial activities was not balanced by the rise of new industries, despite improvements in the electronics sector (AUDIAR, 1985).

Between 1975 and 1982, 10% of the industrial jobs were lost in Rennes District, for two reasons. On the one hand, traditional local firms did not survive an economic crisis, mainly in the building trade and the consumer-goods sector. The two printing houses, including the historic Oberthur, also ran into trouble. The heart of industrial activity in Rennes was gradually disappearing. Second, some firms, which had recently located in the city in the wave of industrial decentralization encouraged by the state, were quick to leave when the first difficulties occurred.

Two types of regeneration strategies were tried during the 1970s. On the one hand, Mayor Freville pursued his former programs and increased their scale. This could be called the *fuite en avant* strategy. On the other hand, the Left Council, elected in 1977, adopted a more cautious approach, waiting for a Socialist government. When the first consequences of the crisis hit Rennes in the mid-1970s, the response of local elites was correspondingly mixed. On the one hand, the Council tried to accelerate development by involving private developers and bankers. On the other, some sections of the university, the unions, and the Left, which enjoyed increasing support during the decade, opposed the new strategy.

By 1971 Mayor Freville was growing old. His support shifted from the original MRP to clear right-wing parties. His apparent heir, a professor of law at the university, was very close to national private developers and was from the *Parti republicain* of President Giscard D'Estaing. A study of Rennes during the 1970s revealed how the new urban planning attempted to adapt Rennes to new capitalist conditions. It focused on prestigious urban programs targeted at the middle classes, in cooperation with national private developers and bankers, creating new infrastructures that were too big for the city, and new commercial developments, including a shopping center close to the city center. Simultaneously, since it was politically attuned to central government, Rennes managed to attract some small firms in electronics and also get public support for area firms with economic problems. By contrast with the earlier period, the new Council's strategy was oriented to the needs of the private sector.

The local perception of Rennes' economic crisis was that it needed to achieve critical size to join the new range of successful French cities. It needed to rapidly expand industries and population and be like Grenoble. Hence, local elites' primary reaction to the crisis was to try making the area more attractive to national and international economic interests. The shift within the Council demonstrated the rise of men close to these interests at the expense of trade-union leaders and voluntary-sector representatives. Government agencies strongly supported these huge development programs since they were in keeping with national strategy at the time.

However, this new strategy generated considerable opposition. First, local shopkeepers and local entrepreneurs, mainly in the building trade thought that their interests were being sacrificed to the profits of larger economic interests—private developers, supermarkets, shopping center developers and owners. Second, what may be called "new social movements" were fast developing within the city. Rennes' expansion had attracted young people from all over Brittany—workers, employees, and lower-middle-class people who went to the university and later worked in schools, hospitals, and local and regional public services. These groups, who lived in the neighborhood built in the 1950s and 1960s, organized to get social amenities, cultural activities, and social provision for their children. These people, who had been very active as students in Leftist groups in 1968, gradually organized within the renovated Socialist Party after 1972 and quickly achieved a rising share of local votes.

These interests vigorously opposed the Council's new strategy. Although most of them were also former members of the progressive Christian Democrat organization, they were on its Left wing. There was also a generation gap between them and the councillors. This new group felt that Rennes' problems were caused by the crisis of international capitalism, and that the strategy of the Council aggravated this by favoring multinational companies over local people.

The local election of 1977 marked both the end of the Freville reign and a major defeat for his right-wing heir. The Left—the Socialist party, the Communist party, and small leftist and regional parties—under the leadership of a young Socialist lecturer in the university, won the election. Their leader, E. Herve, was elected Mayor. A majority of shopkeepers refused to vote for the right-wing candidate, and the Socialist leader attracted their support by campaigning on the defense of local interests, local shops, and a more "humane" urban planning against prestigious urban renewal projects and "big" interests.

Until 1981, in keeping with the traditions of the French Left, Rennes city councillors believed that economic change had to come from the central state. (Barge et al., 1983) Leading councillors explained that an effective local economic policy was impossible without the implementation of three major changes by a Socialist government: (1) the nationalization of key economic actors—banks and corporations; (2) a reform of local finance to introduce a fairer system and real local autonomy; and (3) decentralization reforms that would give real economic powers to the regions. The Communists were even more radical and opposed a priori any local action that could mask the responsibilities of the corporations and of the state in the crisis.

The new leaders of the Council felt that the global crisis of capitalism, whose consequences had reached Rennes, could only be solved by a Socialist strategy at the national level. Because they expected a Left victory in the coming national elections, they were prepared to wait for this national change and implement their Socialist priorities locally. This included developing social and cultural amenities and changing the urban plan, to curtail the more ambitious programs and put some social housing in the nearly completed buildings. This strategy was widely supported by numerous voluntary-sector organizations, trade unions, and local social groups.

However, in this crucial period, 1977-1981, there were three factors that have to be borne in mind. First, the Left did not win in 1978. Second, the crisis deepened and several old local firms collapsed in 1978-1979, increasing unemployment. Third, among the Council, an active minority of Socialist councillors—once part of the progressive Christian Democrat and regionalist tradition, including some union leaders, regionalist activists, and the young generation on the Left—joined the Socialist party. These people were influenced by the ideas of promoting local economic indigenous development, which had developed in rural areas and in some regions, such as Brittany, in the 1970s. For some theorists, local development could even be the alternative to capitalism (*Autrement*, 1983; *POUR*, 1986). Within this tendency, some advocated an interventionist local economic policy with assistance to local small businesses, financial support for firm creation, the development of land and the provision of premises, and research into the local economy. All this was regarded with suspicion by the Council, especially the Communist councillors. However, the pressure of rising unemployment added weight to this approach, and a number of studies and ad hoc responses were set in motion.

STRATEGIES TO REVITALIZE THE LOCAL ECONOMY

In 1981, two facts changed the context of regeneration in Rennes. On the one hand, after the closure of another series of plants, there was increased pressure from the population, the CFDT trade union, and regionalist councillors to develop a local economic strategy. Second, Mitterand became President, the Left won the parliamentary election, and the Mayor of Rennes—from the same Socialist tendency as the Prime Minister P. Mauroy, Mayor of Lille—was appointed a government minister. Finally, local social conditions in Rennes, including the structure of the urban elites, the economic difficulties of the area, and the advent of a Socialist government, formed the background within which a new strategy was worked out. In keeping with the character of the city, it was based on new technology related to the university and a cultural strategy.

SOCIALIST POLICIES AND THE MAYOR-MINISTER FACTOR

The advent of the new government brought three changes: (1) Rennes City Council and Rennes District Council were committed to support a government whose priority was to deal with unemployment; (2) the Mayor became a *Grand Notable* in the French system (Mayor of a regional capital, regional councillor, Minister, former MP) and decided to use his resources to transform Rennes into a European metropolis; (3) decentralization reforms and the process of decentralized regional planning gave Rennes' elites the opportunity to develop and finance their strategy.

A Left government assumed power in 1981, for the first time in 23 years. As a Socialist Minister, the Mayor of Rennes was wholly committed to the success of the socialist experiment, and, inevitably, the Rennes City Council came to the forefront in implementing government schemes. The Mayor of Rennes had been elected at the unusually young age of 34. He was a minister and a colleague of some of the most important *notables* of the French system, including the Prime Minister and the Mayor of Marseilles. He decided that Rennes had to be not just a regional capital—but a real metropolis at the European level. As mayor and minister, he could secure government public investment in the city and exploit the dynamism of local elites. Because of his privileged position within the French system (Dupuy, Meny, & Wright, 1985), he was also able to identify the key areas for the future of cities, and to make sure that Rennes would be among the first to work out one.

During this crucial period in 1982, the Mayor had to prepare the manifesto for the 1983 local election and a *plan de developpement* within the decentralized planning framework. A small team of advisers worked out the major themes for the future economic strategy of the city. The team included a personal adviser and unofficial economic officer and former CCI officer, the head of planning of Rennes District, three leading councillors and committee chairmen, a well-known economist from the university, a regional civil servant close to the Prefect, and the head of the CDC subsidiary, the SEMAEB. The manifesto for the 1977 local election had been mainly designed in contrast by the former Mayor and was highly ideological. By 1983 the Socialist Party had agreed to remain within the EEC and to accept market principles. The old ideology seemed out of date for a young entrepreneurial Mayor attempting to prepare the city for the future.

The third element of the 1981 political changes was the decentralization reforms, the progressive abandoning of traditional interventionist regional policy, and the introduction of decentralized regional planning. Briefly, decentralization reforms essentially provided legitimacy to local economic policies, which were reinforced by centrally sponsored schemes. The process of decentralized planning was supposed to allow a negotiation between regional councils and the regional Prefect about the priorities of the region and public investment. In fact, the two partners were not equal—-only one had finance. Inevitably, most regions did not work out a real plan, but mostly espoused government priorities to secure the allocation of funds for their regions.

This process largely replaced traditional regional policy. It meant that the most powerful of the French *systeme politico-administratif,* the *Grand Notables,* usually Mayors of regional capitals, could secure most public funds for their areas. This was matched by the new priorities of the DATAR, aimed at strengthening French cities in order to prepare for European integration. In Rennes, for instance, the Prefect negotiated the regional Plan with the right-wing regional Council. But the Prefect was appointed by the government, of which Rennes' Mayor was part. The Mayor could refer to him as "my Prefect." In fact, the Rennes' Mayor negotiated specific programs with his government colleagues in Paris and then wrote them into Rennes' development priorities. The Prefect had to accept that this was in Rennes' interest in the regional plan, even if the regional Council did not like it. As a result, more than half of the state-funded projects allocated to Brittany were located in Rennes.

This was the main institutional arrangement that made it possible for Rennes' elites, under the leadership of the Mayor-Minister, to fund key projects and implement their whole strategy. The Prime Minister's first visit to the regions was to Rennes, and he gave special government approval to its strategy, a classic example of how state and local elites interact in the design and implementation of urban policy. It is a two-way process in which the *Notable* plays the key role. It brings local priorities to the center, and central priorities to the locality. The *Notable*'s local authority can suggest experimental programs to the government and also secure central funds for those programs.

Once the local elections had been won by the Left, with its new economic strategy for the area, four changes were made in the preparation of Rennes' *plan de developpement.* First, the intercommunal planning agency was modified to deal with urban and economic strategic planning. Second, within Rennes City Council, a small economic unit was created to deal with assistance to business. Third, the Mayor launched a new body called CODESPAR

(*Comite de Developpement Economique et Social du Pays de Rennes*), which brought together representatives of the business sector, the unions, and local authorities. The chairman was elected every two years but the position had to rotate, that is, a politician first, then a manager or banker, then a trade union leader. This body was to prepare the strategic economic and social planning for the area.

The fourth change was that the CODESPAR, under the leadership of its first chairman, Rennes' Mayor-Minister, took charge of working out a plan for the urban area, which would be included in the regional plan, to get finance from the state for its main programs. The consultation process was very broad. First, an eminent university economist—adviser to the Mayor—and other academics wrote an extensive study of the local economy that identified growth areas. Eight working groups were established to decide policies for each area: agriculture and biotechnology, building trades, electronics and computing, research and development, planning, training, finance, and promotion. Each group had a chairman, and meetings took place over four months. The consultation process was an impressive success. More than 300 people joined regularly one of the working groups—union leaders, representatives of the CCI and business organizations, entrepreneurs in the high-technology sector, training and education officers, councillors, numerous voluntary-sector representatives, managers from the private and the public sectors, and academics. With few exceptions, the urban elites in Rennes worked out the *Plan de Developpement*.

This degree of local mobilization was unique in France. The awareness of the uncertainty surrounding Rennes' future, the chance to attract public investment, the Mayoral leadership, the importance of local networks, and a broad consensus on the future of the city—all contributed to the success of the preparation of the plan. *Le Plan de Developpement du Pays de Rennes* was an interesting mixture of political principles, economic analysis, and specific initiatives, for which financing had to be found by negotiation between local elites and the state. Despite some criticisms from the Communists on the Left and from some business representatives on the right, the document, which provided an overall plan for the economic development of the area, was accepted by all local actors.

The major theme of the document was that job creation in Rennes could be achieved by the mobilization and development of the scientific research capability that had been built up in the previous 20 years. The strategy contained three strands. First, it argued for a permanent organization to mobilize local actors and guide the economic and social development of the area—to make the CODESPAR a permanent body. Second, the area was to be vigorously promoted through coordination of all the *communes*, the CCI, the District, and the university. Third, the economic development of the area was to be pursued in three main sectors: electronics and communications, health and environment, and bio-industry.

The idea was that Rennes' regeneration would occur through scientific and technical research and training. The city had to encourage research and promote the integration of research, training, production, and marketing. It was decided, for example, to create a science park—only two existed at that time in France. The Council was to implement an economic policy including assistance for businesses, the development of local firms and resources, and a program of sites and premises provision. And it was recognized that Rennes had to develop international relations to promote its economic development.

IMPLEMENTING THE STRATEGY

Three elements characterized Rennes' economic strategy. First, through the creation of a science park, it aimed to develop links between research and economic development. Second, the council decided to build on the cultural dynamism of the city, and to integrate new technologies, education and culture. Third, the cultural dynamic and the development of the science park were used for a massive marketing campaign to promote the city as a modern place of culture and intelligence. The whole project was designed to show Rennes as a European city of the future, building on its tradition and history, linking its science park and the old architecture of the city center.

The creation of the science park, baptized Rennes-Atalante, laid the heart of the new strategy. The relations among academics, researchers, and the local authority had always been extremely close. Herve and the Prefect had secured state resources for the park, which was financed through the *Contrat de Plan Regional*—with 4 million francs from Rennes District, 2 MF from the region, 2 MF from the state, 1 MF from the *departement.* A science park society was created, chaired by a respected nonpolitical academic, with a leading councillor and a CCI leading member as vice-chairman. Once again, most of the local actors were involved in it.

The park was built around the university and the public research centers that had been established there in the 1960s. A new neighborhood with quality housing and a pleasant environment had been prepared nearby by the Council. Various networks, events, and collective services were organized, and the District launched futurist premises for high-tech small firms, financed and built in cooperation with the SEMAEB, and the *Societe de Developpement Regional*. The whole venture was very much a public-sector initiative. Once it became successful, the private sector took over in providing premises.

At the same time several plans were developed to improve the image of the city. First, Rennes managed, thanks to the Mayor-Minister's negotiation with the ministry of telecommunication in Paris, to be the first city to get cable television. Second, it created a festival for electronic art, with artists working on computer graphics. This was a national cultural event to promote Rennes as city of culture and technical innovation. In relation to the state, they created a CCSTI, a cultural center for new technology, like *La Villette* in Paris. Successful local exhibitions such as *Rennes au Futur* were launched to examine the consequences of new technology for the inhabitants of cities. They also created a national competition for the best technological change at firm level, which provided Rennes with national media coverage.

The marketing of Rennes was built on the science park, its research center and its young, educated, dynamic population—it has 3,000 people in academic and research jobs and 38,000 students in a population of 300,000. It could simultaneously show the newest technologies in computer graphics and the most successful rock-and-roll festival in France. Innovation and modernization were to be promoted in all sectors. Rennes was one of the first cities to bring cable television into the schools, and to make special provision for poor people. The last measure was endorsed as a central government scheme two years later. New programs in social and educational fields were used to demonstrate that Rennes was a modernist city, a changing and exciting place.

At the same time, a variety of traditional economic policies were pursued: to provide financial assistance, land, premises, and advice to business; to renovate the historic city center and modernize the housing stock; to support a regional cultural festival promoting the identity of the city; to develop the city as a tourist and conference center; to modernize traditional amenities, including a new station for the high-speed trains; to support local firms and jobs.

THE IMPACT OF REGENERATION

To assess the impact of Rennes' regeneration strategy, one has to distinguish among a range of different factors. In terms of its new role in the international or national urban hierarchy, the challenge for Rennes is twofold. It needs to become the main city on the Atlantic side of France, the *Grand-Ouest,* in competition with Bordeaux, Nantes, Caen, and Rouen. Since Rennes and Brittany belong to the periphery of Europe, they both need to avoid marginalization and develop an "Atlantic strategy" with similar regions in other countries, which depends upon the ability of peripheral cities to remain important urban centers on a European scale.

It is still too early to judge the results, but some things can be said. Between 1945 and 1975, Rennes changed from being a quiet provincial city to one of France's most dynamic urban centers. The new urban elites who have been in power in the 1980s have consolidated this transformation. In the 1980s it has become one of the four most dynamic French cities, with Montpellier, Grenoble, and Toulouse. There is some objective evidence to support this claim—the city's rate of research, higher-order services, international relations, new technology firms, the success of the science park, the quality of housing, the employment structure, levels of unemployment. More subjective criteria would also confirm the story—the image of the city, the quality of its cultural life, the quality of the urban environment, its education system. Overall, Rennes during the 1980s has succeeded in promoting an image of "new technology and culture." The success of the science park Rennes-Atalante gave Rennes the basis upon which to sell the other elements of this image of social dynamism and culture. It has 4,500 people working on the site; 10 new firms are created each year; and it may have created about 900 jobs.

Reports from the DATAR place Rennes and Montpellier among the cities with the greatest number of higher-order urban functions in Europe, even though they are not as large as many other French cities. Several magazines have ranked Rennes among the three most dynamic cities in France. Such judgments are fragile, but at least they reflect the success of Rennes' image, which places it among the "modernist and dynamic" cities in France.

If we accept the success of this strategy in the 1980s, it can also be explained by the inability of rival cities, notably Nantes and Bordeaux, to demonstrate similar dynamism despite their size and traditions—460,000 and 640,000 inhabitants. None of them have yet been able to work out a strategy for their urban areas. Their relative conservatism, underlined by the importance of an old local bourgeoisie, has prevented their modernization in recent times. A comparison of the development of Rennes and Nantes would demonstrate the importance of the local social structure in French cities. For instance, throughout the nineteenth century, the bourgeoisie of Nantes, whose wealth had been inherited from the slave trade, opposed the creation of a university. One was only finally created in the 1960s. By contrast, the university has been one of Rennes' major assets during the past 30 years, but especially in the 1980s. Current economic development trends favor innovative cities with a good university, an agreeable and prestigious environment, and cultural activities that attract dynamic social classes. It should be noted, however, that if Nantes and Bordeaux could manage to renovate their dynamism, Rennes might not be able to compete with them.

Indeed, despite its image, innovation levels and sophisticated strategy, Rennes still faces several structural difficulties. The industrial sector remains very weak and has not enjoyed a significant recovery. Although small high-tech firms have significantly increased in number, there is little doubt that Rennes would be in serious trouble if the Citroen plants were to close. Tertiary-sector activities providing more or less qualified jobs have enjoyed significant growth. However, Rennes' economic base lacks the diversity of bigger cities. This could prove an important obstacle in the future since job opportunities for highly qualified people remain very limited, especially in the private sector. This lack of an industrial basis prevents Rennes from reaching a critical size to compete with other French and European cities. Rennes is increasing its international relations efforts to increase its influence. Cooperation with regions in Spain, the UK, and Ireland may help support the Atlantic side of the EEC, but it will take at least 10 years to know whether Rennes will be a significant European metropolis.

A SYMBOL OF REGENERATION

During the 1970s the urban growth policy pursued since the early 1950s in Rennes reached its limits, and the elite consensus became increasingly open to question. The formulation of new redevelopment strategies was affected by both the new social conditions of the city and its increasing economic problems. On the one hand, the old Mayor supported accelerated growth in the city, hoping to adapt Rennes to the new conditions of international capitalism. However, this was a major break with the tradition of thought he represented. As a result, groups from the voluntary sector, new social movements, the hospital, the university, gathered on the Left to organize the opposition and develop a new strategy based on the needs of local people. They formed alliances with local economic interests who opposed the Mayor's projects. The Left won the local elections in 1977 and supported social services and culture in the expectation of a Socialist economic transformation at the national level.

In the 1980s a new political consensus in Rennes was forged around the science park. Its creation demonstrated the leadership of the new Mayor and his ability to mobilize national financial support and the support of local interests for economic development in Rennes. The university, young entrepreneurs, and intellectual elites were particularly important in the preparation of the new strategy for developing the city along the lines of new technologies, culture, and social innovations in, for example, education and training. The innovative strategy plus the ability to rapidly mobilize local actors in support of projects and to work with external partners proved essential for Rennes' development.

It would be wrong to suggest the new strategy did not create conflict; but in the Rennes case, conflict actually stimulated the local dynamic rather than preventing it. Once more, it is essential to stress the importance of the underlying local social conditions, which are a precondition of, for example, the success of a sophisticated local policy, or the creation of a public-private partnership.

Conflict occurred within a regional capital where progressive Christian Democratic values—education, research, culture, and social innovation—were supported by most local groups. The conflict was more about the leadership of the projects. There were two broad camps. On one side were the local authority, the Mayor, the District Council, the planning agency, the CCSTI, cultural organizations, voluntary-sector organizations, the economic development unit from Rennes City Council, the SEMAEB, the university, CFDT, and the hospital. The other consisted of the *department* and its chairman, the region, the Chamber of Commerce and Industries, *Ouest-Atlantique,* business groups, right-wing parties, and some new entrepreneurs. They were united on several projects such as the science park, the *Contrat de Plan,* and the CODESPAR.

Rennes does have to make choices that are faced by other French cities as well as by the Socialist Party. The city's political elites' support has shifted from working-class, public-sector employees, and lower-middle-class people, to public- and private-sector lower-middle-class and middle-class people. The party's strategy has benefited its newer supporters—the lower middle-classes and the city's intellectual and economic elites. The main evidence is its cultural and housing policies, most of which meet the needs of its well educated population. The housing complex and leisure center close to the science park demonstrate the effort to provide well-designed houses and flats with a pleasant environment for people working in the park. It also includes an important part of social housing, but it is for lower-middle-class people, not working-class. In the city center the renovation of the historic quarter and the new housing programs are designed to attract lower-middle-class and middle-class people, as is the heavy emphasis placed on the cultural activities with a budget which now amounts to 15% of City Council expenditures.

On the other hand, the left-wing city council has made some efforts to improve social housing. But overall, lower-middle-class people from the public and increasingly from the private sector, who form the bulk of Rennes' social structure and its political elites, benefit from the modernist strategies. This strategy has raised criticisms from a number of groups. The most severe come from the Communist union, CGT, which stresses the lack of "real industrial locations" in the urban area. But most of the time, the CCI, the ecologists, or the right-wing parties support most of the strategy.

A recent private consultant report confirmed that most of the population enjoyed living in Rennes and being able to take advantage of the social and cultural dynamism of the city. But it also demonstrated how a minority of the population was more and more excluded from the city, with unstable jobs and no relationship to the city center and its activities. Those people were moving gradually to the margins of the city. Significantly, until recently, very few initiatives were taken to support the unemployed. The success of Rennes' modernist strategy has so far been based on the cultural and social dynamism of the city. The development of the trend toward marginalization and exclusion of working-class people could call that success into question.

Rennes, along with Grenoble and Montpellier, which have pursued very similar strategies, is a symbol of urban regeneration in the 1980s in France. The city anticipated the trends of the 1980s and was able to mobilize its dynamic networks of actors to develop and implement a modernist strategy. Rennes is now known as a city of the intelligence, or "thinking network." But to become a European metropolis, it has to take a new step. However, time will tell whether its lack of economic diversity, new social divisions, or employment problems will prove to be difficult obstacles on its path.

REFERENCES

AUDIAR. (1985). *La developpement economique de district du pays de Rennes, 1962-1985.*

Autrement. (1983). Le local dans tous ses etats. (No. 47: Special issue).

Barge, P., et al. (1983). *L'intervantion economique de la commune.* Paris: Syros.

Dupuy, F. (1985). The politico-administrative system of the departement in France. In Meny, Yves, Wright, & Vincent. *Centre-periphery relations in western Europe.* London: Allen & Unwin.

Freville, H. (1977). Un acte de foi. *30 ans au service de la cite* (2 vol.). Rennes: SEPES.

Laborie, J. P., Langumier, J. F., & De Roo, P. (1986). La politique Francaise d'amenagement du territoire de 1950 a 1985. Paris: La Documentation Francaise.

POUR. (1986). Les chantiers de developpement local. (No.106: Special issue).

Teitgen, P-H. (1988). Faites entrer le temoins suivant. Rennes: *Editions Ouest-France.*

5

Hamburg: Crisis Management, Urban Regeneration, and Social Democrats

JENS S. DANGSCHAT
JÜRGEN OSSENBRÜGGE

THE RECENT CHANGES

Hamburg is a port at the mouth of the river Elbe in the north of the Federal Republic of Germany (FRG), some 80 kilometers from the North Sea. It is both a "Land" and a community ("city-state") and, with a population of 1.6 million, the biggest city in the FRG. Traditionally, its economy was based on the harbor, port industries, and international trading. However, after World War II the city became the media center, especially print media, of the FRG and a major insurance center. Hamburg is also the commercial center of Northern Germany and an important transport and communication link to Scandinavia and Eastern Europe. Its current economic structure means that more than 70% of its work force is employed in the service sector (SJDG).

During the 1980s Hamburg faced a very complicated economic situation. It has always been, and remains, a rich city, with the highest proportion of millionaires within the FRG—19.9 per 10,000 inhabitants. But it also has the highest proportion of people on social assistance—930 per 10,000 inhabitants (SJDG, 1988, p. 532). The public sector also faces increasing debt levels. But by 1985 the city had slipped to merely an average position within the FRG. This economic decline created severe unemployment in the early 1980s. However, the position changed during the last part of the decade, and there are now signs that the crisis has been overcome and a structural economic reorganization is under way.

Though this chapter is primarily descriptive, it rests on two assumptions. The first is that the development of urban areas within industrial capitalism can be divided into specific periods with distinctive economic processes and styles of political regulation. In keeping with the long waves of economic development, urban areas in Western economies have been confronted by a general economic crisis since the mid-1970s. The process of economic restructuring of the 1980s has been accompanied by new forms of political regulation (Esser & Hirsch, 1989; Harvey, 1985). In Hamburg's case, this fundamental political change took place in 1983, when the Mayor introduced a new strategy called "Enterprise Hamburg," which redefined the city's urban policies and the role of the public sector in economic development.

The second assumption is that, though the transformation from "Fordism" to "Postfordism" involves a fundamental change in urban development strategies, political elites are still forced to balance economic accumulation and social legitimation policies. We examine Hamburg's economic growth strategies and social investment policies before and after the break in 1983, identifying their basic differences. Particular attention is paid to the way in which structural economic changes have been used to increase the city's capacity to compete with other cities nationally and internationally, and the way that Hamburg's leaders have dealt with social and environmental problems (Ossenbrügge).

The results of the general shift away from distributive justice are discussed in terms of the emergence of the "two-third-one-third-society" in Hamburg. Our discussion begins with a review of demographic and economic trends in the city and a brief discussion of Hamburg's political system.

SOCIAL AND ECONOMIC DEVELOPMENT IN HAMBURG SINCE 1970

DEMOGRAPHIC CHANGE

Hamburg has been losing population for almost a quarter of a century. Between 1964 and 1987 the figure fell by nearly 300,000, from 1.86 million to 1.5 million (Müller, 1989). Since 1970 Hamburg has lost 11.2% of its citizens—an average annual loss of 12,000 people (Hruschka, 1986, p. 37; Stala, 1989). This decline is typical of most of the larger German cities since the beginning of the 1970s, but the loss of population started in Hamburg much earlier. By contrast, Munich, with about 1.27 million—the second-largest city in the FRG—only started to decline in the mid-1980s. The main metropolis of Southern Germany has lost only about 1% of its citizens since 1970, the lowest of all the larger cities in the FRG. These distinctions stem from the two regions' different positions within the development cycle, in particular the delayed industrialization of South Germany.

ECONOMIC STRUCTURE AND DEVELOPMENT

Historically, the regional economy of Hamburg has been dominated by its port-related activities. Trade and transport have been the most important components of economic development since the medieval era. Externally oriented, tertiary economic considerations have always dominated the city's political and investment decisions. Besides these service functions, seaports are always attractive locations for firms dealing with the construction and maintenance of transport and trans-shipment equipment, as well as for those involved in the processing of incoming and outgoing goods. Industries that transform raw materials and upgrade food are typical examples, which can be called "seaward merchandise industries" (Nuhn, 1989).

One basic economic indicator, value added, reveals an overall growth in the regional economy of 1.3% every year from 1970 to 1986 (Stabu, 1988); however, the industrial sector remained constant between 1970 and 1987. This stagnation led to a severe reduction in the industrial labor force, which now constitutes only 60% of the 1970 figure. This deindustrialization is primarily explained by the decrease in shipbuilding, food processing, and petroleum refining: port-related industrial activities facing structural problems and long-term decline. Also, because of inadequate representation in the modern growth sectors, such as chemicals and electronics, Hamburg's production of capital and consumer goods lags behind the national average, too, confirming the city's general crisis in the secondary sector.

Hamburg's job losses in the industrial sector are nearly 10% higher than the average for cities in the FRG with more than 500,000 population. Only some monostructured cities in the Ruhr and Saar areas have experienced greater decline. Today, the proportion of industrial workers is the lowest in Germany with 19%, followed by Frankfurt with 23%. Other main service centers, like Stuttgart and Munich, still have 35% and 27%, respectively, of industrial sector workers.

Despite its deindustrialization, Hamburg's overall urban economy has grown because of the expansion of its tertiary functions. But whereas the growth of service activities has been accompanied by a slight increase in jobs, this is not true for commercial trade and traffic. Here, an enormous increase in productivity has set in, accompanied by the introduction of new technologies that reduce mental and manual labor to a remarkable extent. Once again the main reason is changes in the port. Containerization and new communication technologies have led to a rapid decline in employment, although the port itself remains efficient and competitive on a regional basis. Of course, this restructuring affects all service-sector activities, and, despite the growth of employment since 1970 in private and state services, this is expected to be reversed in the near future. It should also be noted that the expansion of

TABLE 5.1
Firms and Employees in Hamburg 1980 and 1987

	Employees 1970	1987	*Change in %*
Agriculture	3,261	1,811	–44.5
Energy and Water	8,067	3,356	–58.4
Manufacturing	268,657	145,389	–45.9
Petroleum, Chemicals	28,056	14,192	–49.4
Mechanical Engin.	83,062	51,498	–38.0
Electr. Engineering	46,176	28,280	–38.8
Paper and Printing	29,318	10,520	–64.1
Food Processing	39,341	16,879	–57.1
Construction	67,158	51,640	–23.1
Trade	200,700	168,750	–15.7
Transport and Communic.	123,783	96,176	–22.3
Credit and Insurance	46,453	54,179	16.6
Private Services	129,524	254,521	96.5
Cleaning	10,135	32,223	217.9
Education, Leisure	12,561	22,829	81.7
Health Services	13,375	23,719	77.3
Prof. Consulting	34,145	100,944	195.6
Private Associations	20,761	28,353	36.6
State	102,987	131,913	28.1
Total	970,721	936,088	–3.6

SOURCE: Stegen (1989, p. 277).

the service sector has been accompanied by the growth in deregulated work contracts and low-skilled occupations. Finally, fast growing, high value added jobs, like enterprise-related services, form only a very small part of Hamburg's service sector (Table 5.1).

The uneven development of the Hamburg regional economy is well illustrated by its unemployment rate. Two waves of unemployment can be identified. The first, and smaller one began shortly after the first oil price shock in 1974 and ended in 1979. The second one demonstrates an urban labor market crisis that began in 1981 and remains critical, even though the number of unemployed has been falling since 1987.

Hamburg experienced average rates of growth until the end of the 1970s. But in the 1980s the unemployment rate in Hamburg, like that in Bremen, the other major seaport, grew as much as that of Essen, one of the main centers of the old industrialized Ruhr area (Stala, 1989). In contrast, the cities of South Germany are not suffering from such severe unemployment problems. The decline of older industrial cities, the loss of employment in the rural coastal regions, and the growing problems of seaport areas clearly demonstrate the uneven pattern of regional development between North and South Germany (Wieland, 1990).

THE POLITICAL SYSTEM AND THE REGIONAL ELITES

The city of Hamburg is an autonomous state ("Land,") and in this sense is different from the other urban areas in the FRG. The city-state was created in 1937 by the Nazi regime, which wanted to establish an efficient northern growth pole (Gross-Hamburg-Gesetz, cf. Bose & Pahl-Weber, 1986; Hartwich, 1987). Today, the metropolitan area and the economic region of Hamburg is bigger than the Land territory. Though the central city has traditionally been the most successful part, more recently Hamburg's suburban areas have benefited most from the region's general economic success. However, these areas are part of different Laender—Lower Saxony in the south and Schleswig-Holstein in the north. This fragmentation of the region into different administrative units has created a variety of political, planning, and financial problems:

—While Hamburg has been consistently ruled by the Social Democrats, the governments of the neighboring states have changed more often. This has led to problems of administrative coordination and to economic competition between the core of the conurbation and the suburban periphery.

—Different systems of tax incentives and economic support operate which benefit the periphery of the region and place the core area at a disadvantage. This means that potential investors can manipulate political authorities to maximize government subsidies.

—Responsibility for certain regions, which are ecologically sensitive; for example, the Lower Elbe district is divided between three administrative areas. As a result, economic competition has led to the exploitation of natural resources along the Lower Elbe and the coast.

In the city-state of Hamburg, urban economic and planning policy is made at two governmental and administrative levels. The first is the city, which has created large, politically and administratively independent, central departments with sometimes immovable bureaucracies. The second level consists of the seven districts where local policy and planning are administered. This division of responsibility for decisionmaking is complicated and unclear, leading to uncoordinated action and conflict. It also creates uncertainty for potential investors by increasing the time involved in planning and decisionmaking.

All controversial decisions, even if they are specifically local, can be overruled by the state's government (Senat) by evocation. As a result, each minister (Senator) and his administration are very powerful. But the First Lord Mayor, in his function as the Prime Minister of the Land, can only appoint his ministers. Moreover, like the state Hamburg has been governed since World War II, with one exception in the 1950s, by the Social Democrats (SPD), either on their own or in coalition with the Liberals (FDP). This means that party executives are very powerful, since they either are on a rising political career or are being paid off with senior political positions.

The Social Democrats, however, are also internally divided between a left and a more powerful right wing. The First Lord Mayor, Hans-Ulrich Klose (1974-1981), represented the center of the political spectrum but moved to the left during his period of office. His successor, Klaus von Dohnanyi (1981-1988), started as a right-winger and shifted to the liberal-left. Their policies were also affected by the coalitions they had to make, either with the Liberals (FDP) or the Greens (GAL). Since 1988 Hamburg has been ruled by Henning Voscherau, a leading representative of the right wing of the SPD, in coalition with the FDP.

Hamburg's dependence on the harbor, shipyards, shipping, and trade has meant that its elites have been drawn from these sectors and from old merchant families. Traditionally, these elites resisted the industrialization of the city. A "new" elite did emerge after World War II, led by the print media—*Spiegel, Stern, Zeit,* the newspaper and journal editor Springer, the journal editors Gruner & Jahr and Bauer, and the North German broadcasting station (NDR). However, these elites have not yet been accepted into the city's traditional society.

A new generation of politically powerful elites has recently developed with the shift in production to high-tech products—information, electrical, medical, bio-tech, environmental—and the increasing importance of tertiary functions such as mass media, banking and financing, consulting, distribution, and research in new technologies. By contrast, in other cities like Frankfurt and Munich, which had few powerful, traditional elites, the technical elites gained power much earlier and their ideas had more credibility than in Hamburg. This is one of the reasons why Hamburg's economic deterioration was recognized very late. Today both the "old" and "new money" elites are important actors, though traditional elites still have the political advantage.

THE POLITICS OF REGIONAL DECLINE

Although structural economic change caused Hamburg's unemployment crisis at the beginning of the 1980s, the city's urban and regional policies also played an important part in encouraging those general trends. Those policies have to be seen in the context of the basic principles of economic policy that have been traditionally adopted by the SPD. Most importantly these include:

(1) The role of the state is to guarantee the market economy as the most efficient system of regulating the activities of firms and individuals.
(2) Urban policy should intervene in the economy only when there has been market failure.
(3) However, intervention is legitimate in high-risk economic projects to promote research and development in large-scale technologies.

These principles have always guided the ruling Social Democrats (SPD) —proof that the majority belongs to the right wing of the party. However, this liberal ideology has not prevented the ruling party from investing huge amounts of "political money" in key economic projects in an attempt to influence the structural change that has taken place in the economy since the mid-1960s.

MISUNDERSTANDING STRUCTURAL CHANGE

The first attempt to improve the city's economy since the reconstruction after World War II was undertaken in the mid-1960s. This was characterized by the political elites' general failure to understand the changing international economic processes. The most important principle of economic policy during this period was to increase Hamburg's industrial potential by offering new industrial spaces for existing and new firms. The city was seen as under-industrialized and the SPD-Senat, together with the trade unions and the Chamber of Commerce, promoted a strategy to attract industries in the raw materials and producer goods sectors. The development of the Rotterdam Europort seemed a prototype for a new wave of industrialization at locations with access to deep-sea shipping ways. Growth-pole theory provided the context for the so-called Kern-Plan, named after the Minister of Economy, which saw Hamburg as the center of a newly industrialized region on both sides of the river Elbe. New plants of the ferruginous, non-ferruginous and chemical sectors were to be arranged around a chain of nuclear power plants. These ideas were implemented through a port expansion program, major investment by the state-owned electricity generating company "Hamburger Elektrizitatwerke" (HEW), and the provision of physical infrastructure. The neighboring Laender of Niedersachsen and Schleswig-Holstein cooperated with these schemes for their territory on the Lower Elbe because they shared an interest in improving the industrial structure of the region.

The success rate in the new industries was quite variable. The chemical works of Stade (DOW) and Brunsbuettel (Bayer) have became important elements in the local economies. However, the aluminium works in the port expansion area of Hamburg was a financial fiasco because the expected economic benefits failed to arise. However, it took a long time for Hamburg's urban elites to realize that the strategy of relying on heavy industry had failed. Since so much public money was committed to the original industrialization strategy, the city's opportunity to influence structural change in Hamburg's economy was reduced (Ossenbrügge, 1985).

By concentrating on attracting large firms in basic production, Hamburg badly neglected the interests of existing firms. For example, firms wishing to expand had great problems finding locations. Planning decisions took a long time, and between 1965 and 1986 central Hamburg lost more than 114,000 jobs to its suburban zone (Moeller, 1985). There was considerable public discussion of the disadvantages of the city's industrialization strategies after these failures. But Hamburg's political elites simply did not recognize that they were experiencing the beginning of deindustrialization. The only consequence was a shift from bigger to small and medium-size firms. However, the suburbanization of firms continued, but more slowly than at the beginning of the period.

ECONOMIC AND ENVIRONMENTAL PROBLEMS

To cope with Hamburg's unemployment problems, new demand-side instruments were introduced at the beginning of the 1980s under the First Lord Mayor, U. Klose. The SPD's left wing in particular tried to create special instruments to reestablish full employment. A set of labor market policies was used during this period, such as an employment program with a general increase in public investment, and the creation of a second labor market.

The basic instrument in Keynsian regional growth strategies is the attempt to stimulate private investment by public expenditure. As a result, Hamburg's investment budget was expanded during the crisis years of 1980-1983 to the highest levels anywhere in the FRG, financed by additional private loans. The share of loans in relation to the total income of the city-state rose from

12% in 1980 to nearly 20% in 1983. By contrast, at the same time all other German state institutions, especially neighboring governments in the north, reduced their investments and introduced austerity budgets. But "The money Hamburg was spending moved into the neighbouring countries of Schleswig-Holstein and Niedersachsen. That meant that the state of Hamburg had to take over the financial burdens whereas the benefits were localised outside the territory" (SPD, 1986, p. 27). Hamburg's isolated attempt led not to higher investment but to rising debts for the city, which by 1984 exceeded the total budget for the first time. The city's financial situation is still precarious, although the budget problems could be reduced by tax windfalls and the selling of state-owned enterprises. However, this picture might change dramatically if, for example, interest rates rose or the current economic revival came to an end.

Because of the difficulty of measuring the impact of such Keynesian policies generally, after the takeover of the FRG's government by the CDU/FDP in 1982-1983, there was intense discussion in Hamburg to decide whether demand-side policies should be expanded or abandoned. Though it remained partly an open question, it was obvious that neither strategy of the political elites—the industrialization strategy at the beginning and then demand-side policies when unemployment problems took off—was preventing regional decline. On the contrary, misplaced public investments in heavy industries and port-related industries not only had destabilized the city's budget, but also reduced its capacity to explore alternative options.

Furthermore, growing public awareness of the dangers of environmental problems began to highlight the costs of economic development in Hamburg, in particular the expansion of the port and the growth of polluting industries. Indeed, some of Hamburg's environmental problems, such as the contaminated dredging material of the port, the biggest European waste disposal site in Hamburg-Georgswerder, and industrial polluters like the Chemicalwork of Bohringer and the Norddeutsche Affinerie, have become national scandals. These issues explain the electoral success of the Greens (GAL) in Hamburg's parliamentary elections in 1982, which meant that not only did the SPD lose its absolute majority but also that the Christian Democratic Party (CDU) gained support. It also reflected the legitimation crisis within the ruling SPD since the GAL represents the group of German Greens with socialist tendencies, which has attracted support not only for its environmental solutions but also for its social policies to deal with the economic crisis.

The resignation of Hamburg's First Lord Mayor, U. Klose, at the end of 1981 was a major symbolic political event and an indication that the period of Keynesian attempts to address urban problems had come to an end in the city. Shortly afterwards, a similar change happened at the federal government level as the social-liberal coalition was replaced by a conservative-liberal one. However, the SPD in Hamburg was able to recover as a ruling party despite its loss of support and its electoral defeat in early 1982. The successful second election of 1982 of the new SPD candidate von Dohnanyi must be seen against the background of national political events in Bonn. Dohnanyi benefited to a great extent from the sympathy of the voters for the SPD-Chancellor Helmut Schmidt, who, as a citizen of Hamburg, was seen as being betrayed, especially by the liberals. The elections for the city-state in November 1982 ended in a fiasco for the FDP, and the SPD again achieved an absolute majority. Recognizing the significance of his federal electoral windfall, Dohnanyi was determined to use his good fortune to develop a new urban regeneration strategy for Hamburg in 1983.

STRATEGIES FOR MANAGING CHANGE

THE TAKE-OFF: "ENTERPRISE HAMBURG"

In November 1983, two years after becoming First Lord Mayor, Klaus von Dohnanyi gave a speech at the Overseas Club, a forum for the city's traditional elites, in which he paved the way for a new vision of the corporate city of Hamburg. He painted a picture of the city as an enterprise where all members—employee and employer, science and technology, government and opposition, polity and economy—should pull together (cf. Dohnanyi, 1983, p. 3). The speech marked a historic turning point for Hamburg's regeneration strategy. Dohnanyi argued that not only the immediate situation but also long-term trends presented difficulties for the city, in particular the shift of the economic center of gravity of both the FRG and the European Community. In his words, "The historic success of Hamburg is the cause of its current problems" (Dohnanyi, 1983, p. 9).

Dohnanyi called for a policy of aggressive competition with other German and European cities. On the one hand he demanded a strengthening of the city's existing advantages, with better traffic and communication routes for the port and its industries. But he also proposed a shift in the economic goals of the city away from the port, the traditional industries, and trade and toward the media, high-tech industries, and modern services. Moreover, he sought the cooperation of the Technical University and technological groups, as well as increased lobbying on behalf of Hamburg in Bonn and Brussels. He also demanded urban redevelopment to increase the attractiveness of the city by improving its housing, its tourist appeal, its culture, and its sport events and facilities. Last but not least, the traditional SPD urban policy of the redistribution of benefits was to be superseded by a policy of attracting international investment: "The best protection for the socially weak is if Hamburg can be more attractive to the better-off" (Dohnanyi, 1983, p. 21).

Reaction to this major shift in local policy was complex, and split the old trinity of interest, the corporate structure of the right wing of the SPD, the Chamber of Commerce, and the trade unions. The left wing of the SPD recognized the end of the traditional Klose policy and demanded greater emphasis on social justice. The trade unions were afraid of the broad push toward rationalization and the end of subsidies for the declining port industries. For the first time the traditional harbor elite and some of the merchants received official public support from the media as both the oldest and most influential representative of the "old elites."

The results of the controversy became obvious and the political line clearer when, six months later, the Chamber of Commerce called for a new policy for the north of the FRG in general and for Hamburg in particular (Handelskammer, 1984, pp. 19-34). A central theme was the need to reorient economic policy in Hamburg. It promoted the approach labeled "From the Ship to the Chip," the title of an exhibition of the Chamber later that year. The program set a variety of goals, including:

—increased technological competitiveness through cooperation with research departments, especially the new Technical University

—increased investment in all fields of high-tech research and the creation of high-tech parks

—liberalization of the media market—private broadcasting as well as the development of equipment such as fiberglass ropes and enlargement of data transfer

—redevelopment of the area between the town hall and the main station, where the bigger stores were located (City-east)

—creation of a new "entrance" for the city (City-south) and reorganization of the banks of the river Elbe to achieve a better junction between the city and the port

—revitalization of the inner city by providing apartments of a higher standard for affluent customers of the shopping arcades, restaurants, and cultural events

—support of cultural events, fairs, international congresses, and sports events to attract international tourists and boost local tourism

These proposals by the Chamber of Commerce became the regeneration strategy for Hamburg, in the context of Dohnanyi's notion of the "enterprise city".

ACCUMULATION POLICIES

PLANNING TECHNOLOGICAL INNOVATIONS

The creation of a milieu that would stimulate technological innovation formed the basis of Hamburg's new industrialization policy. The roots of this policy lay with the decision to establish a Technical University in Hamburg in the late 1970s. Its location in the southern district of Hamburg was intended to achieve two goals: to close the economic gap between the northern and southern parts of Hamburg, and to develop the economy of the southern region in general and support the small and medium-size firms by cooperation in research and production. Other supporting policies were introduced during the first half of the 1980s:

—In 1980 the Hamburg Technological Institute (HIT) was added to the Technical University to establish an institutional arrangement for the private application of university knowledge.

—The Technological Advisory Centre (TBZ) was created in 1985 to improve the transfer of technologies between the universities and the economy. The primary aim was to organize contacts between scientists and private firms informally as well as in major fairs.

—The top-down approach to new technologies was complemented by a bottom-up approach in the field of environmental technologies. The Chamber of Craftsmen's Trade created a Centre of Energy, Water and Environment (ZEWU) in 1985 to train small entrepreneurs and skilled workers in the field of environmental protection.

—Technological projects, ranging from artificial intelligence to port-related transport communication systems, to new shipbuilding technologies, were also pursued.

—The environmental challenge was transformed into an economic growth strategy in the first environmental action program in 1984, by stressing the export potential of urban environmental industries.

FROM SHIP TO CHIP

Port-related activities faced both a crisis of production in the shipyards and a recession of world trade in the late 1970s and early 1980s. But with the recovery of international transport and the growth of containerized goods, there has been substantial public investment to change the role of the port. "From Shipyards to a Full-Service Distribution Center" is the new slogan of the enterprising city. The basic component of the port modernization has been the development of an electronic communication system (DAKOSY) that links all port-related enterprises, such as freight forwarders, packers, and shipping companies, with all their international contracts. This accelerates trans-shipment and distribution and makes possible just-in-time production. Furthermore, new container terminals and the infrastructure for storage and warehouses have been constructed.

While the port was a symbol of crisis in the early 1980s, it has become a strategic component in recent accumulation policies. The term "Eurogate" illustrates the efforts of the economic development institutions to improve the port infrastructure so it can handle the increased flow of transport from Eastern Europe, the Far East, and Scandinavia into the EC and vice versa. A new development plan for the port has also been laid. Since Hamburg celebrated the 800th anniversary of its port in 1989, there were plenty of opportunities for the "Harbor mafia" to make sure that the future shape of the port would be just as the ruling elite wants it to be. The corporate power of the port administration, private business, and union leaders clearly expects to put the new proposals into practice. There remains some doubt about this since earlier port expansion caused severe conflict among the authorities, the ecology groups, and local people who were expelled by the new port development. A new round of criticism and conflict can be expected in the early 1990s.

MANAGING ACCUMULATION POLICIES

To carry out the new economic policy, the government of Hamburg, the Chamber of Commerce, and the 10 most important banks established, in July 1985, a limited company, the Hamburg Business Development Corporation (HWF), to both attract new firms and fund and support existing ones. The chief of the 25-person team was a former member of the managing board of the AEG trust—a troubleshooter in the fields of plant construction, shipbuilding, and high technology. A headhunter hired him for the DM300,000 job—the second such appointment made in the FRG.

His aim was to slow down industrial suburbanization, strengthen existing branches rather than attract new ones where it was felt that cities in the south of the FRG had the advantage over those in the north. Focal points of the strategy were to be the media, information technology, electrotechnology, aviation industries, medical technologies, biotechnology, harbor-oriented services, and environmental technologies.

In its first four years HWF attracted 267 firms from Scandinavia, the UK, Switzerland, Japan, Taiwan, Korea, and the USA. They contributed to the creation of 10,300 additional jobs and attracted investments of about two billion DM—mostly within the construction sector—which fueled a tremendous inner-city construction boom. This company symbolizes the typical German public-nonprofit-private partnership.

FROM ACCUMULATION TO LEGITIMATION

Besides its effort to create a high-tech urban milieu and attract new investment, which we have called the politics of accumulation, the SPD strategy for the reconstruction of Hamburg has a social dimension as well. The first element of this politics of legitimation was the city's attempt to address the crisis of the shipyard industry. It was recognized that intervention was necessary in 1983 when the workers of HDW went on strike and occupied the yard for several days. But the Senate did not develop a strategy to qualify the remaining workers for alternative jobs until 1986. The 1986 Economic Action Programme identified strategies such as wage support for participating workers, new educational facilities to improve skills in metal branches, and the coordination of all initiatives to improve human capital using modern information systems. The rationale for this effort is the growing demand for skilled workers in the city despite the high unemployment rate. These economic incentives have been transformed into a social policy by devising a qualification program for former workers in the shipbuilding industry. In this way public investment to improve regional human capital has been combined with a social defense strategy.

Apart from this response to the special problems of the urban labor market, the centerpiece of the politics of economic legitimation is the "Second Labor Market." The basic idea of this initiative is an attempt to redirect financial aid and public transfers, like unemployment benefits and social assistance. Money which was formerly paid directly to the unemployed was used to finance employment opportunities in the public sector. Since 1983 the Federal Institute for Employment (BFA) has sponsored programs which offered jobs for two years as long as they did not compete with private-sector jobs or regular public employment. Federal money was primarily distributed among the Laender according to the number of long-term unemployed and the rate of unemployment, with the Laender obliged to contribute at least 20% of the total costs of the program.

Hamburg used this instrument especially intensively between 1983 and 1985. The creation of a Second Labor Market was a sequel to the demand-side policies of the early 1980s, but was more flexible. In this way, earlier expectations of economic spin-off were replaced by direct subsidies for the unemployed. According to the new liberal economic ideology, intervention in economic and structural change is no longer required.

The Second Labor Market has proved a successful way of addressing problems of legitimation without endangering competitive regional policies. Those who got jobs through the program serve as a small-scale multiplier because they were active in the fields of social services and environmental rehabilitation. In this way the SPD successfully integrated both of the most urgent problems—unemployment and environmental deterioration—into its urban hegemonic system.

THE ROLE OF SPATIAL PLANNING

All Hamburg's renewal and improvement programs of inner-city areas before 1983 were designed to slow down the suburbanization which had led to population losses of more than 15,000 per year in the early 1970s. In 1971 national legislation provided for the renewal of the old, run-down inner-city housing financed by the federal state, the Laender, and communities. Hamburg also financed its own modernization programs and shifted its attention from the entire city to the ward level.

A shift of focus from the expansion of the city to a rehabilitation of the existing stock can be seen in the urban development plan of 1980. However, the price of the successful reduction of suburbanization was a loss of cheap housing in the inner city. Moreover, the inner city became attractive for real estate agents and speculators because housing costs are relatively low compared with other German cities. Among the six most important cities of the FRG, the journal *Capital* (1989) found the highest profits in Hamburg's inner city, and the main reason for this, besides the delayed price rises, was the improvement of older housing stock and environment by the city.

The planning programs produced growing protest from those citizens who were affected by them, and this sometimes included district governments. There were protests against the location of a new musical theater in a lower-class housing area, the location of a major journal publisher next to a workers' area on the bank of the Elbe, the location of a sport dome, and housing policies that encouraged displacement, rent rises, speculative vacancies, and conversion into condominiums. The protests coalesced into a general criticism of urban planning in the inner districts.

One element of the revitalization of the CBD, particularly the western part, was the construction of eight shopping malls between 1978 and 1983, which was financed by insurance companies. These activities have since spread across much of the city. The construction boom became obvious by the mid-1980s with the establishment of the City-South complex, the intensification of land use by City-West and the northern bank of the river Elbe. The predominant postmodernist architecture style has become a symbol of the economic power of the growing parts of the city's economy. Planners regard these activities as the most important guarantee of a positive future for the city. By 1993 between 3.5 and 4 billion DM will be invested in new construction mainly in the CBD.

In fact, the city now lacks a master plan for urban development. The current plan has lasted 16 years, with only a couple of changes in response to small-scale problems. There is neither a plan for the traffic system nor a development plan to protect the environment. These gaps are even criticized by some of the city's senior administrators. But planning has been cut to pave the way for economic growth, even when it contradicts existing plans or legislation.

In this context the city administration has two diametrically opposed options. It can continue to oppose suburbanization by boosting gentrification, or it can keep or enlarge the stock of cheap and public housing in the inner-city areas and give the lower social strata a "right of immobility." Which of the two will be pursued is unclear. The first approach was endorsed by the Minister of Building, the city's chief architect, heavily supported by the powerful right-wing of the SPD and the Chamber of Commerce and was carried out during the 1980s.

THE STORY OF THE TWO-THIRDS SOCIETY

There is no doubt that revitalization in Hamburg will continue—boosted by a variety of forces. The question remains, who will profit and who will lose? What is the social price of economic revitalization? One clear result is the distinction between the two-thirds better-offs and the one-third losers.

For example, while 151,000 inhabitants (9.5%) received social assistance in 1987, the figure rose in one year to 167,000 (10.5%). The city spent 665m DM for social assistance in 1989. The number of people receiving housing subsidies increased from 64,000 (in 1981) to 84,000 (in 1987) with a parallel rise of payment from 93m DM to 179m DM. And finally, between 1988 and 1989 the number of homeless increased from 8,000 to 26,000.

The differences in rent burden, the ratio of income to rents, clearly show striking differences. The Hamburg average is 22%. In the micro-census of 1985 (Stala, 1985), 31,600 low-income households (those below DM 800) had an average rent burden of 63.5% in 1985. By contrast, the average rent burden was 14.5% for those 29,400 people with incomes above DM 4,000. The rent burdens are also high for single-person households, 26.2%; householders below 25 years of age, 27.5%; and those in newer dwellings, 30.0%. Housing costs in Hamburg have increased and will continue to increase in the future.

These patterns of social differentiation—boosted by the market and supported by urban planning—naturally translate into residential segregation. On the one hand, renewal, modernization, speculative vacancies, and the shift from rental to owner-occupation attack renters by questioning their "right of immobility," and finally displace them. On the other hand, those who need public housing will be concentrated in the high-rise housing estates of the 1960s and 1970s, since this will constitute more than 80% of the dramatically decreasing public stock by the turn of the century.

Households who are slightly above these limits will be moved to refuges of the "traditional poor"—areas of postwar reconstruction. These are badly equipped, small dwellings in neighborhoods that have suffered disinvestment by the public and private sectors. The "new poor"—young adults, sometimes well-educated, single-parent households—and the pioneers of gentrification will prefer to stay in inner-city areas. But these areas are also in demand from the gentrifiers, creating conflicts between them and the pioneers as well as between the pioneers and the public authorities who are the agents of the revitalization.

The city of Hamburg is on its way to being divided into four quarters (Dangschat & Krueger, 1986). They are:

(1) Areas where average affluent German families live, and the traditional peripheral housing areas of the upper classes. These are by far the largest areas.
(2) The housing areas of the "traditional poor" and lower classes, either the big housing estates or, shifting from there and from the inner city, the areas built in the 1950s.
(3) The traditional inner-city, upper-middle-class housing areas of the bourgeoisie of the turn of the century which, as much as the neighboring areas, are becoming increasingly gentrified.
(4) The inner-city areas of conflict, mainly renewal or modernization areas of former workers' neighborhoods. Pioneerism of the 1970s and 1980s has paved the way for a further upgrading of these (Dangschat & Friedrichs, forthcoming).

Even if the economic problems of Hamburg's last economic crisis seem to be resolved, social problems will inevitably occur in the future. They will arise from increasing socio-spatial inequalities, particularly from those with the ability and readiness to fight for their interests. A senior administrator has observed: "It is surprising that Hamburg has overcome the enormous structural change without severe social conflicts." If the city continues to follow the policies laid down during the 1980s solely devoted to economic growth, he may find he spoke too soon.

THE NECESSITY FOR A BALANCE

The social and economic problems faced by Hamburg since the 1970s mirror the crisis faced by comparable old industrialized cities. Our analysis has revealed the initial inability of the economic and political elites to respond to the structural changes taking place in the international economy. They were slow to recognize the crisis they faced. Existing urban economic policy was outdated and it emerged as misguided expensive public investment. This only perpetuated the crisis as declining industries were subsidized for far too long.

As we have seen, new strategies were adopted during the early 1980s, and today politicians, administrators, and private firms emphasize that the city has passed through the valley of failure and is showing the first signs of a revitalization. What definitely has changed is the psychology of economy. Politicians of both ruling parties attempt to encourage economic success with their policies. The HWF praises its acquisition activities, the administration of the Building Minister points to its far-reaching planning. Last but not least, local managers emphasize their flexibility.

The political elite of the ruling SPD has shown that the party is able to find a strategy beyond Keynesianism. In both the politics of accumulation and legitimation, new approaches and flexible policy instruments have been introduced that have preserved their support among local voters. But despite this, the economic, political, social, and spatial changes that began during the early 1980s have increased inequality with urban areas and between social classes.

Though it is an open question whether increasing inequality and growing competition between desirable and affordable housing will lead to open conflict, the challenge of distributional justice will only increase. The successful continuation of "Enterprise Hamburg" demands social and environmental transformation as well as the development of strategies that will gain broad public approval.

REFERENCES

Bose, M., & Pahl-Weber, E. (1986). Regional- und landesplanung im Hamburger planungsraum bis zum "Groß-Hamburg-Gesetz" 1937 [Regional and territorial planning in Hamburg until the implementation of the "Big-Hamburg-Law" in 1937]. In M. Bose, M. Holtmann, D. Machule, E. Pahl-Weber, & D. Schubert. . . . *ein neues Hamburg entsteht. Planen und bauen von 1933-1945* [. . . a new Hamburg is developing. Planning and construction from 1933 to 1945] (pp. 9-15). Hamburg: VSA.

Dangschat, J. S., & Friedrichs, J. (Eds.). (forthcoming). *Die wieder-belebung innenstadtnaher wohnviertel—gentrification in Hamburg* [The revitalisation of central city's neighbourhoods—gentrification in Hamburg]. Opladen: Westdeutscher Verlag.

Dangschat, J. S., & Krueger, T. (1986). Hamburg im sued-nord-gefaelle [Hamburg's position within the south-north-gradient]. In J. Friedrichs, H. Haessermann, & W. Siebel (Eds.). *Sued-nord-gefaelle in der Bundesrepublik?* [South-north-gradient within the FRG?] (pp. 188-213). Opladen: Westdeutscher Verlag.

Dohnanyi, K. von, (1984). Unternehmen Hamburg [Enterprise Hamburg]. In *Ueberseeclub Hamburg* (Ed.). (pp. 3-28). Hamburg: Jahresband 1982/83.

Esser, J., & Hirsch, J. (1989). The crisis of fordism and the dimension of a "postfordism" regional and urban structure. *International Journal of Urban and Regional Research, 13*, 417-437.

Handelskammer (Handelskammer Hamburg, Chamber of Commerce) (1984). *Herausforderung für den norden—zur diskussion um das wirtschaftliche sued-nord-gefaelle* [Challenge for the north—on the discussion of the economic south-north-gradient]. Hamburg.

Harvey, D. (1985). *The urbanization of capital.* Oxford: Blackwell.

Hartwich, H-H. (1987). *Freie und Hansestadt Hamburg: Die zukunft des stadtstaates* [Free and Hanseatic city of Hamburg: The future of the city-state]. Hamburg: Landeszentrale fuer politische Bildung.

Hruschka, E. (1986, February). Bevölkerungsbewegung in Hamburg und seinem um-land 1967/75 bis 1984: im trend der entwicklung der großstädte? [Demographic changes in Hamburg and its region between 1967/75 and 1984: Parallels the general trend of bigger cities?]. *Hamburg in Zahlen*, 36-40.

Immobilien mit sicherem gewinn. Jetzt stadtwohnungen und haeuser kaufen [Real estate with a secure profit. Buying condominiums and houses now]. (1989, February). *Capital*, 95-109.

Moeller, I. (1985). *Hamburg.* Stuttgart: Enke.

Müller, G. (1989, July). Erste strukturdaten aus der volkszaehlung 1987 [First Structural Figures of the 1987 Census]. *Hamburg in Zahlen*, 200-203.

Nuhn, H. (1989). Deindustrialization and problems of revitalisation in the Hamburg Port Area. Paper presented to the British-German Geographers Symposium 1989 *Deindustrialisation and new industrialisation in Britain and Germany.* Hull and Cambridge.

Ossenbrügge, J. (1985). Industrialisierung der peripherie: Aus-wirkungen neuer standorte auf die regionalentwicklung in Nord-westdeutschland [Industrialisation of the Periphery. Effects of new locations on regional development in northwestern Germany]. In P. Rieckmann (Ed.). *Alternative hafen- und küstenkonferenz* [Alternative Conference on Harbour and Costal Problems] (pp. 21-35). Hamburg.

Ossenbrügge, J. (1988). Regional restructuring and the ecological welfare state—spatial impacts of enviromental protection in west Germany. *Geographische Zeitschrift, 76* (2), 78-96.

SPD. (1986). *Arbeit und umwelt für Hamburg* [Work and environment for Hamburg]. Beschaeftigungsorientierte Alternativen zur Standort-politik. Arbeitskreis Wirtschaftspolitik der SPD Hamburg-Eims-buettel. Hamburg.

SJDG. (1988). *Statistisches jahrbuch Deutscher gemeinden* [Statistical yearbook of the communities of the FRG]. Koeln: Der Deutsche Staedtetag (Ed.).

Stabu (Statistisches Bundesamt, Statistical Office of the FRG) (Ed.). (1988). *Lange reihen zur wirtschaftsentwicklung* [Longtime figures on economic development]. Wiesbaden.

Stala (Statistisches Landesamt der Freien und Hansestadt Hamburg) (Ed.). (1985). *Mikrozensus 1985* [Micro census, 1985]. Hamburg.

Stala (Statistisches Landesamt der Freien und Hansestadt Hamburg). (1989). Hamburger zahlenspiegel [Statistical supplement]. *Hamburg in Zahlen, 11*, 358-363.

Stegen, H-E. (1989, September). Hamburgs wirtschaftsstruktur. Erste ergebnisse der arbeitsstaettenzaehlung 1987 [Economic structure of Hamburg. First results of the census from 1987]. *Hamburg in Zahlen*, 276-283.

Wieland, K. (1990). *Regionale krisenentwicklung in den wirtschaftsraeumen Hamburg und Ruhrgebiet, traditionelle ueberwindungsstrategien und alternative loesungsansaetze* [Regional development of crisis in the areas of Hamburg and Ruhr, traditional solutions to overcome and alternative solutions]. Frankfurt: Lang.

Zweite karriere [Second career]. (1987). *Wirtschaftswoche, 36*, 67-68.

Part II

Preserving the Gains

6

Regeneration in Glasgow: Stability, Collaboration, and Inequity

ROBIN BOYLE

GLASGOW'S CHANGING IMAGE

"Glasgow: a city in collapse" was how John Gretton titled a *New Society* article published in 1972. Some 18 years on, a cursory scan of newspapers and other media would quickly suggest that his prediction of hopelessness was deeply flawed. By the mid-1980s Glasgow was attracting a very different press. Commonplace was the hyperbole found in *Time* magazine: "The City that Refused to Die—Once slum-ridden Glasgow renews it center—and its spirit." And following Glasgow's designation as European City of Culture in 1990, the headlines became positively euphoric: "Glasgow isn't Paris, but . . . " wrote Bill Bryson in the *New York Times Magazine* of July 1989. Even the English press waxed lyrical as Glasgow celebrated the launch of the European City of Culture. McKie (1989), writing in the *Observer*, described the new Glasgow as "a city shedding its skin like a rare reptile."

In sharp contrast to Gretton's image of a gray, hard, violent city, Glasgow in the late twentieth century has manufactured a new set of urban symbols. In place of heavy engineering, dismal tenements, and sectarian division, the largest city in Scotland (population 700,000) now sells itself as a specialist center in the service economy, offering a quality of urban life attractive to international capital and tourist alike. On a scale unprecedented in the UK, Glasgow appreciates the importance of image in the 1990s, and in particular the psychological, political, and economic benefits of manipulating urban imagery. Just as Glasgow was used by Gretton and many others as the example of urban despair, so the city has learned to use the media—in print, on film, on video—to flaunt its emerging reputation as a model of the postindustrial city.

This chapter endeavors to ask a series of questions about Glasgow's regeneration, placing contemporary change into the thematic structure of the book as a whole. What was the socioeconomic context that first led to long-run decline, then changed, relatively quickly, to offer the city the possibility of a postindustrial future? And what then was the machinery of urban adaptation: how was this change effected, was there a grand plan, who was involved, what resources were employed? What then is the outcome of urban regeneration: who has benefited and what has changed? Indeed, is there any substance to Glasgow's renaissance, or is it simply a patina of improvement that obscures much deeper problems of postindustrial urban decline?

THE CONTEXT OF CHANGE

Postindustrial change in Glasgow is all the more stark because of the city's rich industrial heritage. As in many northwestern British cities in the nineteenth century, Glasgow's extensive private, corporate, and civic wealth was based initially on a combination of successful trading with North America, mainly in tobacco, followed by phases of sophisticated and profitable textile manufacture—first linen, then cotton (Gibb, 1983). Utilizing good water quality and the proximity of hydraulic power, by 1800 cotton spinning was the staple Scottish industry, giving rise to numerous settlements in the Clyde valley, with the emerging city of Glasgow becoming a regional "command and control center" with worldwide trading connections. The wealth created out of textiles and the attendant development of a scientific and physical infrastructure in the region was put to good effect as the city emerged as one of the heavy engineering capitals of the world. Indeed, by the early twentieth century, Glasgow would boast that it had claim to the title "Second City of the Empire" (Oakley, 1975).

The comparative advantage of the region was largely due to the local reserves of coal and ironstone. In the districts surrounding the city, iron smelting and its associated technology became the bedrock of the economy. There is an argument, however, that the high proportion of output exported to England (as much as 70%) sowed the first seeds of industrial weakness as the Scottish industry failed to experience the full regenerative growth that could have resulted from local manufacture and trading. Nonetheless, iron production and the skill-base of the work force quickly led to the development of new processes and products: especially iron ships on the Clyde and railway manufacture in Glasgow.

TABLE 6.1
Population Change in Glasgow

Date	*Population*		
1801	77,000		
1831	202,426		
1851	300,000		
1871	477,732		
1891	658,073		
1911	785,000		
1939	1,128,473		(Estimate)
1961	1,065,017	}	
1971	898,848	}	Change 1961-1981: –32.8%
1981	774,086	}	
1986	725,130		
1990	694,817		(Projection)
1995	662,299		Change 1986-1995: –8.66%

SOURCE: Glasgow District Council (1989b), Cunnison and Gilfillan (1958)

It was for shipbuilding that Glasgow gained its position in the world economy. Between 1870 and 1914 Scottish shipyards were responsible for one-third of the total British output, and in the peak year of 1913, the 39 Clydeside yards launched 750,000 tons of shipping, nearly one-fifth of the world's tonnage (Oakley, 1975). Some 60,000 men were employed in shipbuilding with another 40,000 in ancillary and related industries, giving Glasgow, at the turn of the century, the lowest unemployment rate of any British industrial region. Glasgow was booming. New industrial processes were started: sewing machine manufacture at the Singer Plant in Clydebank, motor vehicles (Arrol-Johston, Albion, and Argyll), specialized steel production (Colvilles), and integrated manufacturing, including armaments, as in Beardmore's Parkhead foundry complex in the city (Keating, 1988).

The population of the city expanded accordingly, rising from 77,000 in 1801 to 300,000 in 1851, reaching 785,000 before the First World War (see Table 6.1). And the city grew to accommodate the influx of workers from the countryside, the Highlands, and Ireland. This resulted in high-density housing that would eventually give rise to slum conditions on a scale unprecedented in British cities. This alternative urban legacy would eventually determine much of the shape and direction of public policy in the twentieth century.

TABLE 6.2
Employment Change (%) 1971–1983 by Industrial Category

Category	%	Category	%
Food, Drink, and Tobacco:	–29.7	Paper and Publishing:	–42.1
Metal Manufacture:	–77.2	Construction:	–13.1
Mechanical Engineering:	–59.6	Gas, Water, Elect.:	–27.4
Electrical Engineering:	–40.3	Transport and Comm.:	–35.8
Instrument Engineering:	+17.4	Distributive Trades:	+30.2
Shipbuilding/Maritime:	–38.6	Insurance/Bank/Finance:	+42.8
Vehicles:	–25.7	Professional Services:	+12.3
Clothing and Footwear:	–53.3	Miscellaneous Services:	+18.5
		Public Administration:	+8.4

SOURCE: Lever and Mather (1986)

Glasgow's economic decline was less abrupt than its earlier ascendancy, but the urban impact was just as telling. The post-1918 slump and the depression in the 1930s weakened the economic base of the city, exposing the vulnerability of its heavy industries to stagnation in world trade. Unemployment peaked at 30% in 1930. The nadir in shipbuilding was reached in 1933 when output sank to 7% of its 1913 peak. Not surprisingly, the steel industry was also affected. Of even greater concern was a failure of the Scottish economy to capture its share of modern industry in the inter-war years. Moreover, Keating (1988) records that the "entrepreneurial culture of the Victorian period all but disappeared with the attention of the remaining big industrialists concentrated on cartelisation, protection and monopolisation."

Sidney Checkland's famous study of Glasgow used the metaphor of the *Upas Tree* to illustrate what was happening to a city in decline:

> The Upas tree of heavy engineering killed everything that sought to grow under its branches . . . now the Upas tree, so long ailing, was decaying, its limbs falling away one by one. Not only had it been inimical to other growths, it had, by an inversion of its condition before 1914, brought about limitation of its own performance [Checkland, 1981].

And it is widely held that the decline of the city was not assisted by a failure to develop new industrial relationships. Instead, Clydeside and Glasgow became synonymous with union militancy and defensiveness, protecting outmoded working practices. Equally at fault were the employers who clung to a belief in autocratic control with an antiquated conception of workplace management. Alan Massie's summary of decline makes for painful reading:

> Clyde-built had been a synonym for quality; now the Clyde seemed to represent all that was worst in British industry: restrictive practices; obsolete ideologies;

TABLE 6.3
Employees by Employment Categories, 1978–1986

	1978	*%*	*1981*	*%*	*1984*	*%*	*1986*	*%*
Manufacturing	116,400	29	91,000	25	77,000	22	61,000	19
Construction	30,450	7	26,200	7	25,500	7	24,100	7
Services	257,150	64	246,800	71	245,000	71	240,900	74
Total	404,000		364,000		374,500		326,000	

SOURCE: Glasgow District Council (1989b)

> weak, unimaginative and increasingly desperate management; inefficiency; over-manning; short-sightedness; the absence of any intelligent sense of direction [Massie, 1989, p. 102].

The decline of the city's economy continued in the 1960s and 1970s. Both shipbuilding and metal manufacture recorded dramatic falls in employment, and by 1980 there were only two operational shipyards on the Clyde, employing less than 13,000 people. As shown in Table 6.2, this pattern was repeated across almost all manufacturing sectors. Between 1971 and 1983, manufacturing employment in the city fell by 45%. By the end of the 1970s, Glasgow had effectively lost its industrial base (Lever & Mather, 1986).

Critically, Glasgow failed to make good this decline by creating sufficient jobs in the service sector. While the proportion of jobs in services increased, the number of people employed in this sector actually fell (see Table 6.3). However, this shift in the economic base of the city was the trigger for postindustrial policy development that will be fully examined in the following sections.

The human cost of the economic transformation was recorded in the level of unemployment in the city. Taking one snapshot, in July 1986, when the UK was experiencing substantial economic growth, some 77,000 people in Glasgow were counted as unemployed, an overall rate of 21.5%. Following national trends, from the mid-1980s, unemployment in the city began to fall, with the figure at June 1989 standing at 55,000, or 16.2% of the registered work force. Nevertheless, this improvement masked higher levels of unemployment in selected parts of the city, with 10 Wards recording overall unemployment levels in excess of 25%, particularly in the peripheral housing estates that encircle the city. Hard core unemployment was particularly significant, with 25% of those out of work being unemployed for more than three years (GDC, 1989a).

STRATEGIES FOR CHANGE

The theme common to journalists and others writing about Glasgow in the 1980s has been the search for that crucial date when the renaissance of the city was supposed to have started. There is a similar discussion on the single most important event that triggered urban improvement. Other commentators seek to identify lead urban actors who are commonly accepted as being the catalysts of urban change. Despite the plethora of articles, documentaries, and opinions on Glasgow, perhaps the only certain conclusion from media and academic attention on the city is that change cannot be attributed to any single cause or decision. Nor is it possible to find a convenient single date in the calendar as the starting point of urban improvement. Similarly, the search for the key politician or city official or leading businessman tends to exaggerate the personality factor as an ingredient of urban change. Rather, the improvements that have attracted attention to Glasgow are illustrative of a much more complex and possibly fortuitous combination of chance, historical circumstances, political, economic and social context, national policies, and local endeavor.

As an aid to understanding change in the city, Glasgow's recent history will be explored through four interconnected phases: *Renewal, Rehabilitation, Regeneration,* and *Reinvestment.* Although they reflect chronological order, as will become apparent, there is considerable overlap across these four periods. To assist analysis, as far as possible a common set of criteria will be applied to each phase to link the process of urban evaluation in Glasgow over the past 30 years.

URBAN RENEWAL 1955-1970

To understand the demographic, social, and spatial structure of Glasgow in the late twentieth century, it is necessary to first appreciate the mechanisms and impacts of urban renewal in the 1950s and 1960s. The contemporary city was effectively reshaped by four public plans: the advisory Clyde Valley Regional Plan, published in 1946; the first city Development Plan, approved in 1954; its quinquennial review in 1960; and the Greater Glasgow Transportation Study of 1967 (Wannop, 1986b). The emphasis of these plans was on slum clearance, public-sector housing construction, inner-city comprehensive renewal, and motorway construction—in essence, the modernization of Glasgow.

Albeit with varying significance, all four plans illustrate the influence on the restructuring of the city by central government. Indeed, the history of postwar Glasgow is one of competition and conflict between the original Glasgow Corporation and the Scottish Office in Edinburgh. This conflict came to a head over the issue of population dispersal from the city. Through the sub-regional plan, central government sought to move some 500,000 people from Glasgow's slums to new towns being built across central Scotland. The city politicians and their officials were vigorously opposed to this scale of decentralization and instead advocated a combination of high-density housing development on peripheral sites acquired by the Corporation in the 1920s and 1930s and the redevelopment of inner-city slum areas. The outcome was a compromise between the city and central government with the product being considerable population dispersal to the New Towns and other towns in Scotland, the building of four major residential extension areas within the city boundary (to be later labeled "peripheral estates"), and the beginning of an ambitious comprehensive urban renewal program.

What is significant about this period was the dominance of the public sector. Government agencies, including the New Town Development Corporations, worked with the city in pushing population dispersal and house construction. But it was with the authority, funding, and variable support from the Scottish Office that the Corporation of Glasgow, firmly in the control of the Labour Party since 1933, implemented the municipalization of housing in the city. The lead actors in the city were predominantly in the public domain, with the most important political positions being concerned with land, planning, and housing. Yet in contrast to the postwar rebuilding of other British cities, it is difficult to identify key figures. Considering the earlier political history of Glasgow, the impact of socialist leaders on the city is not immediately apparent. Nonetheless, having gained control of the city in 1933, the Labour Party, led by (Sir) Patrick Dollan, was able to make use of Scottish reforms introduced earlier by the Labour "Clydesider" MPs. John Wheatley's housing legislation in the 1920s and Tom Johnston's administrative reforms while Churchill's Scottish Secretary during the Second World War (Smout, 1986) certainly left a legacy of planned renewal, for it was during his tenure in the Scottish Office that the Clyde Valley Plan was devised and partially implemented (Wannop, 1986a).

These politicians were responsible for the massive public housing program that transformed both the tenure structure and the physical appearance of the city. Adapting extensive housing legislation enacted throughout the 1930s and 1940s, the city authorities sought to provide low rental housing, ridding the city of chronic overcrowding and clearing the notorious slum areas such as the Gorbals, where densities in excess of 400 persons to the acre were commonplace (Gibb, 1983). Initially, the city built housing estates ranging from low to medium density, with an adequate provision of local facilities. But the scale of Glasgow's housing problem and the system of housing finance served to replace quality with quantity, resulting in major housing projects located on the edge of the city, increasingly higher densities, and a general poverty of social and community facilities. While an understandable response to the problems of the day, this housing policy served to sow the seeds of a social problem that would reemerge 20 years later.

Extensive municipal rehousing in the city also left Glasgow with an extensive legacy of the modern movement, reflecting the social and political priorities of the period. Although influenced by similar renewal schemes in North America, this phase of urban renewal in the city created an unusual morphology. While strict planning controls prevented high-rise commercial redevelopment in the city center, an enthusiasm for multistory housing led to more than 300 public-sector high-rise blocks being built across the city, many in the Comprehensive Development Areas (CDAs) designated in a zone surrounding the city core (Jephcott, 1971). But like the housing on the periphery of the city, these "villages in the sky" were quick to turn sour, leaving numbers of families isolated and embittered with little chance of being rehoused in traditional tenement areas or in the popular low-rise schemes.

The importance of the public hegemony in Glasgow was not simply confined to housing but was also reflected in the transport planning of the city. The Bruce (1945) plan for the city, devised in the immediate postwar period, had been developed around a complex road system. Little progress was made until the mid-1960s, when first the 1965 Highway Plan and then the 1967 Transportation Study advocated a major urban motorway network for the city. Using overly optimistic population projections and exaggerated claims for car ownership, the city obtained the necessary Scottish Office support to implement an ambitious program of three concentric motorway rings around the city, linking into the emerging interregional motorway network to the south and east. In all, 167 miles of motorway and 117 miles of new expressway were planned, at a cost of (then) £390 million. It was argued that the location and scale of building could be justified because construction would run in tandem with the inner-city redevelopment programs.

As with many ambitious road schemes of the 1960s, only a fraction of the urban network was to be completed. Nevertheless, the construction of the M8 Motorway, forming a loop across the river Clyde and around the city center, gave the city rapid access to the Scottish/UK motorway network. Moreover, in subsequent years the city was to widely advertise the benefits—real and illusory—of a mere 15-minute drive from the city center to the "Shuttle" to London and the international markets.

Ironically, it was this investment in infrastructure during the 1960s that was later cited as one of the mechanisms employed by the city to effect the transposition from an industrial to a postindustrial center. First, improved communications, travel time to the airport, and access to motorway sites were to figure prominently in the marketing material issued in the early 1980s. Second, a number of traditional city center land uses, the fruit market being a case in point, relocated close to the motorway, releasing land that, 15 years later, would become valuable redevelopment sites in the city center. Third, and linked to this change in city center land use, a case has been made that the demolition of older housing areas, the clearance of sites, and the resulting "motorway blight" reduced land values in the inner core to such a level that they then offered the potential of low-cost private urban renewal. This theme will be reexamined below.

The product of urban renewal in the city was profound. In statistical terms, the city was transformed. From a high point of 1.12 million in 1939, the population fell to 889,500 by 1971. Land, planning, and especially housing policy effectively shaped a "municipal city." This was most obvious in housing, so much so that by 1970 some 54% of all homes in the city were owned and managed by public landlords. In effect, the middle class retreated to the suburbs. Between 1960 and 1970 public housing authorities built 42,000 houses in the city. Over the same period, only 1,500 houses were added to the private-sector stock. And between 1952 and 1971 employment in the city fell by 18%, the human cost of the closure of traditional industry in the city.

URBAN REHABILITATION 1968-1980

On the morning of January 5, 1968, Glasgow was battered by a freak hurricane, given the somewhat unspectacular title of "Low Q." The whole city was affected, but it was the older, traditional tenement housing stock, most of it still in private ownership, that took the brunt of the devastation. More than 10,000 roofs were damaged through a combination of the wind removing elderly "nail-sick" slates, water penetration, and serious structural damage caused by falling chimney-heads. The significance of this event was that the city administration (at that point still controlled by Labour), with the reluctant support of a Labour Secretary of State, had, as a matter of urgent priority, to yet formulate a program of action that would improve, not remove, private-sector housing in the city. Therefore, Low Q had the demonstrable effect of highlighting the importance of Glasgow's tenement stock, the shortsightedness of a myopic housing policy that was, at times, antipathetic toward the private market, and the urgent need to help individual home owners with some form of grant aid. Furthermore, the hurricane galvanized political and public opinion. There was £7.8 million allocated to the city within a month of the disaster; it was estimated that between £25 and £30 million would be required simply to make good the storm damage.

Therefore, the city had to formulate a positive rather than a negative response to the private housing sector. Rather than ignore or alienate the home owner, grant aid was made available that would not only repair the tenements in the short run but would also commit central and local government to a program of capital investment in its older, private housing stock. This sea change in Glasgow's housing policy, and the overall effect that had on the city, was to reappear at different stages and through different channels throughout the next two decades.

With hindsight, the combination of changed perceptions, new attitudes, and the necessary stimuli coincided with changing legislation that provided the vehicle for urban change. Building on legislative change at the end of the 1960s, the Housing Act of 1974 and the Housing (Scotland) Act of the same year offered the city the means to shift its housing policy from clearance and renewal to that of rehabilitation of the older, largely private housing stock. First, the city could select Housing Action Areas where they could offer grant aid to individual home owners and, second, they could designate Housing Associations as the agents for individual house and community renewal. Maclennan (1983) has argued that the scale of low-income owner-occupation and the level of rental income in the private rented sector militated against a "private" strategy for tenement improvement. However, the use of Community-Based Housing Associations, funded by the UK Housing Corporation, was ideally suited for the rehabilitation of the lower end of Glasgow's private housing market. The CBHA also proved politically acceptable and, moreover, the Housing Corporation was a convenient mechanism for increasing central government investment in the city during a period of public-sector retrenchment.

The benefits to the city of the CBHAs can be measured in a number of ways. In contrast to other UK cities, the improvement to working-class tenement housing was targeted at, and received by, the lower-income groups in the city without the detriments of either gentrification or the wholesale displacement of urban neighborhoods. Keating (1988) makes the point that the Associations energized community self-help and local cooperation across the city. Moreover, the local control that they exercised integrated the community into area improvement as well as simply housing renewal. Indeed, this model of community partnership, with the CBHAs working closely with the Council's area improvement strategy, has been identified as a positive feature of renewal in the city. And the program has had a physical impact as well. Much of the very fine stone-cleaning and other environmental improvements in the older parts of the city offer visual evidence of the benefit to the city of this approach to housing improvement.

At a different scale, the implementation of the 1973 Local Government (Scotland) Act in May of 1975 provided an important policy framework for Glasgow's gradual improvement. Although far from universally popular, and bitterly contested by the then-Corporation of Glasgow and by the surrounding Conservative-held suburban authorities, the creation of a two-tier system of local government was eventually seen by city politicians as being to their advantage. The scale of resources attracted by Strathclyde Regional Council, their policy commitments to Areas of Multiple Deprivation, "at risk" groups, and the problems of unemployment have all served to focus attention on the city (Strathclyde Regional Council, no date), probably giving Glasgow a greater share of the available resources. Moreover, the strategic framework developed by the Regional Council was also to benefit the city. Through their transport, planning, and industrial development powers, the Regional Council drew attention to the problems of the conurbation and promoted a number of the policy recommendations first raised in the West Central Scotland Plan (Wannop, 1986b). The policy focus on the problems of the city itself—especially the designation of Urban Renewal Priority Areas and the identification of Economic Initiative Areas—afforded important political support for the city even if the resources were not quite so forthcoming.

There is little evidence that the private sector in the city played a significant part in the shaping of policy in this phase. They obviously took a role in supporting housing rehabilitation, but in general the business community was absorbed by the desperate economy of the city. Instead, the lead agencies were in the public domain: the Housing Corporation at the national level, the Community-Based Housing Associations at the local level, and the newly established District and Regional Councils.

URBAN REGENERATION 1975-1985

There are four key elements to this phase in the contemporary restructuring of the city: first, the critical importance to the city of a government agency, the Scottish Development Agency, and its funding of environmental change and economic development; second, the remarkable change in housing and planning policies resulting in a shift from municipal domination to mixed tenure; third, a continuation and extension of inner-city community renewal, often supported by the SDA; and fourth, the activities of the city and the private sector in changing the external image of the city, rebuilding community confidence in Glasgow, and beginning the process of urban marketing.

As has so often been the case in the modern history of the city, external influences played a key role in the economic and physical regeneration of the city. Much has already been said on the subject (Keating, 1988; Keating & Boyle, 1986; Lever & Moore, 1986), but it is difficult to overestimate the importance to the new Glasgow of the Scottish Development Agency. Charged with the task of regenerating the Scottish economy, the SDA was formed in 1975 by the UK Labour government in response to intense political pressure from the Scottish Nationalist Party. Lewis Robertson, the SDA's first Chief Executive, was certainly reluctant to commit the SDA to urban regeneration. Pressure from the Scottish Office and the political complexion of the Board—not least the first Chairman, Sir William Gray, former Lord Provost of the city—was crucial in moving the SDA into urban policy. Its principal role was the management of a commitment to a renewal program "GEAR" (Glasgow Eastern Area Renewal) Project (Donnison & Middleton, 1987; Keating & Boyle, 1986). The level of Agency expenditure through the GEAR project (£78 million by 1986), as well as extensive investment in Land Renewal (later Land Engineering) in other parts of the city, was a key factor in translating political objectives into tangible urban change.

Despite an initial reluctance by city and regional politicians to accept the SDA as a major player in urban regeneration, the financial resources and the political weight delivered by the SDA quickly dispelled any thoughts of obstruction. This places the role of the SDA in Glasgow in sharp contrast to the initial response of city authorities in England who resented the imposition of the Urban Development Corporations. Instead of hostility, the SDA was seen as the catalyst for urban regeneration, offering the city the resources necessary to implement plans and projects first conceived during periods of financial stringency.

By the late 1970s the success of the Housing Association movement in the city began to have an impact on the monolithic Housing Management Department of the city council. Younger politicians and their officials sought an alternative to the crude municipalization strategies of the "old guard." When the Labour Party lost control of the city between 1977 and 1980, the Conservative Administration, its officials, and the younger Labour councillors in opposition found the opportunity to examine a new approach to the city's housing policies. On regaining control of the city council in 1980, Labour announced its "Alternative Strategy," aimed at maximizing the use of existing resources and encouraging the flow of private resources into housing while, as Keating (1988) argues, "avoiding the wholesale privatisation favoured by the Conservatives." To reduce the impact of falling capital allocations and, at the same time, to widen housing choice in the city, the Strategy sought to harness the resources of the private sector. Grant aid (up to £5,100 per unit) to house builders tackling difficult conversion schemes and the disposal of buildings and property owned by the District Council was an important factor in attracting the private developers into some of the most difficult housing areas. In April 1984 the Labour Group reversed the Strategy begun in 1980, placing a moratorium on the further disposal of council-owned land to private house builders. This was a reaction to the problems encountered over the demolition of Hutchie-E, a notorious prefabricated housing estate in the Gorbals. Consequently, for two years the building rate of new private houses fell until, in late 1986, the Council introduced a new "Pilot Programme," using the release of council-owned land to stimulate new private housing development.

The City's Planning Department also began to relax certain constraints on private development within the city center, coinciding with the policy shift in the Housing Department and the availability of grant aid and the introduction of LEG-UP support (Local Enterprise Grant for Urban Projects) from the SDA. Indeed, as early as 1981, the draft Central Area Local Plan identified the Merchant City as a non-statutory "Special Project Area." A formal Local Plan was eventually published in 1986, but the city planners were able to bypass this bureaucratic mechanism, promoting instead development through a combination of selectively releasing land (the city owned 40% of the vacant sites in the Merchant City), selling or giving away empty property (60% Council ownership), and packaging available grants and loans for the private sector. By 1988 some 1,964 houses had been completed, 612 were in the pipeline, and a further 1,120 were planned or under negotiation.

This phase in the city's adaptation to a postindustrial world will be best remembered for the Council's vigorous approach to urban marketing. In 1983 Glasgow District Council launched a campaign of municipal marketing and unabashed self-promotion. With the support of city businesses and institutional interests, the slogan "Glasgow's Miles Better" became the ubiquitous symbol of a changed city. Every conceivable means of publicity was employed—ranging from full-page advertisements in the color supplements of the national press to elaborate publicity stunts in London—in an attempt to persuade a skeptical public, at home and abroad, that Glasgow had indeed found itself a new role in postindustrial Britain. No longer was this to be the city of the Gorbals slum, religious bigotry, and uncontrolled street violence, but instead a city of culture, rediscovered affluence, and personal opportunity.

The campaign was conceived by the then-Lord Provost (Mayor) Michael Kelly and designed and managed by John Struthers, senior partner in a local advertising agency. Kelly had been so impressed by the success of the "I Love New York" campaign that he convinced others that the city should promote a similar venture. Combining the slogan "Glasgow's Miles Better" with the well-known "Mr Man" cartoon character, Struthers designed a series of promotions using local, regional, and national media. Building on initiatives such as car stickers, municipal merchandising, and local advertising, in 1984 the council took its campaign to London, selling Glasgow on the sides of double-decker buses, on taxis, on the London Underground, and in the mainline railway stations. This coincided with a series of advertisements placed in the national (UK) press and, more significantly, in a number of the specialized journals targeted at the international business community.

The campaign needs to be seen as more than simply civic hype, but was built on the idea of the cultural city. The early 1980s were a period when a series of individual projects and events matured at the same time, producing an important critical mass of cultural activity. The opening of a new art gallery to house the eccentric but internationally renowned Burrell Collection was undoubtedly the catalyst that drew the different components together. This, together with the opening of a new Royal College of Music and Drama, firm proposals to construct a new concert hall, the emergence of Scottish filmmaking talent, and the rediscovery of a new Glasgow "style," not least in the visual arts, served to convince the UK government that Glasgow should be nominated as European City of Culture for 1990. Glasgow's cultural resources: home of Scottish Opera and the National Orchestra, the Citizens Theatre, and the largest municipal art collection outside London, became part of the marketing literature, joining references to the city parks, access to outdoor recreation, and proximity to the Scottish Highlands, promoting the city for tourism, for inward investors, for business.

Looking across the whole city and taking account of different sectors reveals that the lead agency in this phase of urban regeneration was undoubtedly the Scottish Development Agency. But it is to misunderstand change in the city to suggest that it offered strategic leadership. More often, it added its resources and political weight to existing projects and initiatives. Indeed, what was distinctive during much of the early 1980s was the lack of any formal planning framework in the city. Land use planning in the city was particularly weak, with plans driven by resource allocation and feasible project implementation.

As for private-sector involvement, the business community was initially reluctant to play a central role in urban regeneration. With little hope of achieving lasting political power, individual conservatives and members of the Chamber of Commerce and other industrialists sought to establish specific linkages with the public sector, particularly in the field of property development. Albeit slow to respond, the private sector also saw the positive advantages of the "Miles Better" campaign and eventually rallied to provide further funds for the Kelly/Struthers initiative. Nonetheless, this significant phase of urban change was largely organized, led, and funded by the state.

The impact of urban regeneration on the city between 1975 and the mid-1980s was tangible, highly visible yet curiously remote from the reality of urban decline. Miles better, perhaps, but Glasgow still faced problems of chronic unemployment (in excess of 45% in certain city wards), poverty (1986 Grieve Report on housing in the city recorded that 70% of households had a gross principal household income of less that £5,200, [GDC, 1986]), and business failure. Even in the initiatives areas such as GEAR, the impact of urban regeneration was limited. One study found that policy resulted in a mere 445 jobs being created between 1976 and 1982, of which 187 went to local people (McArthur, 1987). With 4,000 unemployed in the area, urban regeneration can be a dangerously exaggerated term.

URBAN REINVESTMENT 1985-

The focus of this most recent phase in Glasgow's restructuring reaffirms the conclusion reached in the preceding section. According to a series of studies in the early 1980s, the public sector focused its attention on the city center with clear policies targeted at developing a service-based economy, to be implemented through increased support for new commercial development, the attraction of external capital, and urban tourism. In support of physical change, agencies in the city developed a sophisticated approach to place and property marketing. This strategy crucially involved reviving the influence of the business community.

Following an extensive economic study of Glasgow, the Scottish Development Agency launched Glasgow Action in 1985 (Boyle, 1989). Consultants' reports (McKinsey, 1984) recommended that public action, channeled through the SDA, should seek to strengthen Glasgow's role as a major service-sector city and that such a strategy should be led by a private organization. The SDA then approached a group of prominent city businessmen as potential board members, and asked Sir Norman Macfarlane, already an SDA board member and chairman of a successful local printing and packaging firm, to lead the board of Glasgow Action. A range of other businessmen was selected from across a spectrum of locally owned companies to join the board. It is noticeable, however, that all were well connected in the Glasgow and Scottish business community, having numerous interlocking directorships (particularly in Scottish financial institutions), membership in the local Chamber of Commerce, and the CBI. In the classical style of the "partnership" model, city government was also represented, with two senior Labour politicians from Glasgow District Council and Strathclyde Regional Council joining as board members. But the publications, the activities, and the style of management clearly suggest that leadership, control, and direction were to be firmly located in the private sector.

Yet Glasgow Action is a public body. It is solely funded by the SDA, its staff are employees of the Agency, its offices are within the complex that houses the SDA, and its activities are recorded in the SDA's Annual Reports. The Director of Glasgow Action is even listed alongside the Regional Directors of the SDA.

Essentially, Glasgow Action seeks to use its private-sector profile and its public-sector resources to function as a development catalyst, stimulating ideas, projects and schemes, building the appropriate connections among different actors in the process of urban development and promotion. Hence the *modus operandi* of Glasgow Action represents a radical departure from the substance and style of the earlier phases of urban regeneration, when the SDA played a key role in directly funding its Area Projects and operating in relatively close harmony with a series of public partners (Keating & Boyle, 1986). In the 1990s Glasgow Action operates in such a way as to promote private activity, using the SDA and local government where and when required. Glasgow, it should be pointed out, was not the only city where the SDA sought to champion the private sector. Other similar initiatives were launched in the mid-1980s, notably the Aberdeen Beyond 2000 Initiative. During the same period, the SDA also became a major sponsor and promoter of local enterprise trusts throughout Scotland, working closely with the business community in specific urban locations.

The significance of the progression of the SDA's activities in the realm of supporting private leadership in the cities is only now, in early 1990, being fully recognized. In 1990 the Scottish Development Agency will merge with the Training Agency in Scotland to form *Scottish Enterprise,* which has the remit "to provide a clear vision and direction to the design and delivery of enterprise development, training, and environmental projects and services in Scotland" (LECU, 1989).

This new Agency will operate through a series of 12 "Local Enterprise Companies" (LEC), each run by boards where at least two-thirds of the directors, including the Chairman, must be senior figures from the private sector. There may be minority representation from local government, but it is certainly not mandatory. The LEC will have a broad set of duties, primarily concerned with training and business development. Nonetheless, of the 11 activities listed in the Scottish Office manual (LECU, 1989), 4 functions relate directly to urban regeneration: (1) designing and ensuring the implementation of urban renewal projects and program, (2) securing property development, (3) securing land renewal and environmental improvement projects, and (4) managing industrial property.

At present (early 1990) local consortia of business interests have been invited to submit bids to the Scottish Office to form the Local Enterprise Company for areas determined in advance. Resources of up to £100,000 for "development funding" will then be made available to the successful applicants in order to produce a detailed business plan for the LEC. Returning to Glasgow, it is not surprising that Glasgow Action has played a central role in the putting together of the business consortium for the Glasgow LEC. Moreover, Sir Norman Macfarlane is to be Chairman of the Board. The LEC bid has also been supported by staff who currently work for the Glasgow Directorate of the SDA, hence it is likely that there will be a close correlation between the new Company and the existing policies and programs of the SDA's current activities in the city. Although final details have yet to be determined, it is likely that the LEC will operate in Glasgow with an initial annual budget from the Department of Employment (Training) and the Scottish Office of upwards of £50 million.

In essence, therefore, urban economic development, urban regeneration, business development, and the whole range of training services will be administered through the private-sector-led LEC. It will enter into contracts with the government to supply these services, then subcontract to a range of private and public companies and agencies to deliver the appropriate services. Hence, the implications for urban management, the future of local government, the direction of urban regeneration, and the long-run purpose of Scottish urban policy will be profound. Moreover, the measurement of performance—of both the LEC and its subcontractors—is likely to concentrate more on value-for-money than on the economic or social objectives of urban policy. Only time will tell, but the early indications suggest that it will be the criterion of business that will determine significant parts of Scotland's and Glasgow's urban policy in the 1990s.

Continuing the trends of the 1980s, there is little formal opposition from local government to the creation of the Enterprise Company. Control of Glasgow's urban policy had already effectively passed to the SDA's Glasgow Division while restructuring in the City Council had further weakened the role of the Planning Department. Economic policy was transferred to the Chief Executive's Department, but other than fulfilling a useful information function, it is difficult to identify its impact on the city's economy.

In addition to its involvement through Glasgow Action, an indigenous private sector began to play an important role in stimulating and delivering property development. Just as the Victorian city was developed by a number of small property companies (the Laurie Brothers, the Glasgow Building

Company, Gilmorehill Land, Stobcross Proprietors, and others), the contemporary conversion process is Glasgow has been promoted by small, Glasgow-based developers. Companies such as Jim Oliver's "Windex," the Lafferty Company (now bankrupt), Barrie Clapham's "Crudential Holdings," Andy Doolan's "Kantel," and others have successfully read the changing property market and skillfully utilized the resources on offer from the SDA and the City Council.

Some of these small developers have evolved into larger, more ambitious companies capable of undertaking major renewal schemes that have then attracted financial support from some of the major institutions, in particular the banks. Once almost solely concerned with housing, companies such as "Classical House" and "Windex" are now tackling complex multiuse developments, commonly in city center locations. Hence, there is an emerging progression by the new Glasgow developers that begins with grant-aided residential conversions in the tertiary market, then evolves into retail, industrial, and particularly office development in more profitable locations. And this progression has, to an extent, been assisted by the activities of the SDA and its support of Glasgow Action.

It is an exaggeration to suggest, however, that the property boom in the city is being developed predominantly by local businesses. The principal players are London- or Edinburgh-based financial institutions working with a combination of local and national (UK) developers. This interest in the city by UK institutions has resulted in international finance being attracted to Glasgow, often in partnership with multinational developers. An example of this trend is Glasgow and Oriental Developments Ltd (GOD), a company formed in 1986 by Kumagai Gumi UK, the property development and investment subsidiary of the Japanese parent company, Bellhouse and Joseph Investments Ltd of London, and the SDA. GOD's first development is being constructed (early 1990) on a waterfront site at the Broomielaw, with the first phase providing 276,000 sq. ft. of office accommodation and 59 flats. But it is the funding of this project that offers further evidence of the attraction of capital to the city. With security underwriting from the Mitsubishi Bank (Japanese) and the Clydesdale Bank (now under Australian ownership), GOD refinanced the development, selling six packages to other Japanese banks (including the Tokyo Trust) and the Toronto-Dominion Bank of Canada.

The expansion of tourism, the 1988 Glasgow Garden Festival (largely funded and managed by the SDA), and the successful capture of the UK nomination for European City of Culture, 1990, are all part of a process of vigorous municipal marketing. These and other initiatives can be analyzed in terms of the city's positioning itself to attract mobile capital. This explains the consensus between government and the private sector, the important role played by Glasgow Action, and the capitulation to the Local Enterprise Company. Local government support for privatism is not simply the preferred approach in Glasgow; it is seen as an essential requirement for civic survival (Barkenov, Boyle & Rich, 1989).

THE PRODUCTS OF URBAN CHANGE

It should come as no surprise that the process of regeneration and the more recent commercial reinvestments in Glasgow produced mixed fortunes for its citizens. Not only are there numerous examples throughout the twentieth century of the benefits and disadvantages of urban change, the very history of Glasgow is one of uneven development. Looking back to the industrial heyday of the city, to the so-called Second City of the Empire, in essence there were two nineteenth-century cities. The history of Victorian Glasgow is, as Massie (1989) records, "one of triumph, of an expanding economy, a city growing ever richer and more splendid, erecting magnificent public and domestic buildings, a city rich in high culture, notable for piety and philanthropy . . . but it is also a story of degradation and misery, of the fierce exploitation of man by man." While the economic and social baseline is different, the fruits of urban renaissance in the late twentieth century are similarly unequal.

Looking first to the advantages of modern Glasgow, it is important and necessary to begin with the renewal of self-confidence in the city. As much as anything this has been the product of substantial physical improvement, resulting in the generation of a new civic pride. As with the Great Exhibitions in 1888 and 1901, and the Empire Exhibition of 1938, the 1988 Garden Festival was as much as anything a celebration of confidence in the city. Glaswegians like to be proud of their city.

The four phases of urban change discussed above have produced, both by accident and by design, population stabilization and a reduction in net outmigration, resulting in a measure of community renewal. And the impact of policy in the 1970s and 1980s has been to engender, albeit slowly, a belief in community self-determination, even in some of the poorest peripheral estates. Moreover, change in housing policy and the activities of new local developers has resulted in a wider range of housing tenures. Most of the publicity is attracted to city-center change, but the impact of increasing tenure choice across the municipal stock has been to the city's advantage.

As an outcome of contemporary reinvestment in Glasgow, the city center is capturing external capital. The product is usually measured in terms of floor space, but the payoff in primary and secondary employment, additional capital investment, and further image enhancement serves to consolidate a positive trend. As with other cities, the service economy has been further enhanced by the development of a tourist industry, resulting in new facilities and added employment. Accurate statistical measurement of this change has yet to emerge, but on the basis of the impact of employment generated through the arts, the increase in tourism is likely to result in further economic advantage for the city.

If contemporary writers are in agreement that Glasgow has improved, they are all equally adamant that the benefits are selective. The dual city thesis has been well rehearsed elsewhere (Mulgan, 1989), but in Glasgow the concept is all too apparent. First, the immobility of capital across the city means that private investment rarely reaches beyond the commercial core. Indeed, analysis of property development reveals the spatial selectivity of the investment industry (Ellis, 1986; Erdman, 1990). There has been little investment beyond the principal business district that has not received extensive public subsidy. Moreover, the level of nonresidential private investment in renewal areas is very limited. In GEAR, the 10 years of public investment resulted in a leverage ratio of 1:0.5; for every pound of public finance, the private sector committed 50 pence.

Moreover, very small proportions of private finance have found their way to the core problem areas: the peripheral estates and other desolate public-sector housing schemes in the city. A recent initiative by central government that focuses on peripheral estates, "New Life for Urban Scotland" (Scottish Office, 1988), simply underscores the poverty found in these areas and freely admits that private funds are unlikely to flow into communities such as Castlemilk and Easterhouse. Hence, as predicted (Keating & Boyle, 1986), the dual economic structure of the city has resulted in a dual urban policy: robust urban economic policy supporting city-center commercial renewal opposed to underfunded, fragmented social policy in the housing schemes.

The peripheral estates and other "Areas of Priority Treatment" are the other side of "Culture Capital." These areas contain concentrations of the most deprived households in the city: areas where "income differentials, the growth of unemployment, inadequate housing and the failure of many services to provide for the most needy" (Strathclyde Regional Council, no date) have combined to disadvantage so many. There are more children per family than in the city as a whole, their health is poorer, less than 2% of the heads of households are in professional or managerial occupations, and 40% of those fortunate to be in work are in low-paid jobs. The level of poverty is most clearly revealed by the statistic that two-thirds of households in the Easterhouse and Drumchapel received housing benefits and that less than 15% of families had access to a car (CES, 1985).

It is misguided to suggest that politicians and policymakers have ignored the peripheral estates. On the contrary, the past 15 years have seen numerous initiatives and well-meaning programs aimed at the worst areas in the schemes. But despite the best efforts of community workers and others, these initiatives have too often failed to attract sufficient political support to offer resource discrimination on a scale reflecting the severity of the problems. Limited and disjointed finance, professional division, and inadequate control all served to minimize the impact of public effort. And, despite the rhetoric of the organizers, the impact of European City of Culture, 1990, is difficult to find in the desolate schemes of Possil and Barmulloch.

Yet, while the concept of the divided city finds considerable intellectual support, there is little political or community-based opposition to the way in which current policy has evolved. Ian Jack, writing in the *Independent,* concluded that urban renewal has resulted in "little more than a brightly embroidered shroud," and that the city's economy is based on "self-advertisement and history." Where then is the translation of this argument into discontent or political action? With the exception of one short collection of poems, stories, and slogans (McLay, 1988), there is very little published criticism about contemporary Glasgow. The answer to the silence lies perhaps in two completely different explanations.

First, Glasgow's poor are simply no longer part of the city. In economic, political, or social terms, their alienation from urban life is now almost complete. Years of isolation bred hopelessness; the response of the very poorest has been to withdraw into a personal world of survival, far removed from city center renewal, waterfront housing, and new shopping centers. If the City of Culture means anything, it is little more than an obscure TV program that happens to be filmed in Glasgow. Alternatively, Glasgow is more than just an example of a postindustrial city. Its residents—of all classes—have embraced the consumer culture of the late twentieth century, while at the same time retaining the city's distinctive character. Just as they cheered the launch of the Queens on the Clyde, then signed on for Unemployment Benefit, Glaswegians in 1990 have learned to adapt to a new, perhaps superficial, economic order, but are equally determined to be proud of their new city. "Miles Better" than what? Who cares!

PROTECTING THE GAINS

Running throughout this overview of leadership and change in Glasgow is the argument that the length of time involved in effecting substantive urban regeneration is considerable. In Glasgow's case, certain key infrastructure decisions in the early 1960s and significant events toward the end of that decade played a crucial role in setting the city on a course that would, eventually, lead to modest improvements in the 1990s. Moreover, it is on the basis of decades of public investment that the private sector now sees the advantages of taking the lead. Without such long-run underwriting, it is unlikely that commercial interests would seek to get involved.

Hence, UK and Scottish government has shaped modern Glasgow as much as any other individual force. The key economic decisions affecting Glasgow, covering, for example, mergers policy, the shape of the steel industry, the scale of office relocation, to name but three, have all been determined by national government and, on occasion, by the Secretary of State in Edinburgh. Similarly, a stream of postwar Secretaries have consistently made detailed policy decisions about Glasgow population movement, house construction, and transport planning that have effectively restructured the city and, in turn, limited the power and influence of "local" politicians. Moreover, the reorganization of local government in the 1970s, dividing responsibility for the city between two councils, served to assist the Scottish Office's "Divide and Rule" policy and certainly enhanced the role for the SDA in the city (Keating, 1988).

Without exaggerating its importance, the Scottish "community" has played a key role in determining how the city changed in the late twentieth century. As Kellas (1975), Keating and Midwinter (1983), and Keating and Boyle (1986) have all argued, the closeness of political, professional, and commercial institutions has produced a distinctive form of policy-making and delivery in Scotland, one that has had a significant impact on Glasgow. The Glasgow Eastern Area Renewal project (see Donnison & Middleton, 1988) was a case in point, where a relatively traditional urban renewal program was (successfully) delivered through a combination of eight public agencies, coordinated by the Scottish Office, chaired by Labour and Conservative Ministers alike.

Many writers share the belief that the long Labour hegemony has, ultimately, been to the city's advantage. What councillors may lack in the way of policy thrust they make up for in terms of putting Glasgow first. Despite years of mistrust, "business interests recognise that the Labour administration is a natural part of Glasgow life and are the more ready to accept that they must work with it. At the same time, the administration has learned that the recovery of prosperity depends on preferring collaboration to divisive rhetoric" (Massie, 1989, p. 120).

Finally, if there is a single lesson to be taken from the Glasgow case study, it is that of consistency: in terms of general objectives for the city, in terms of an overall urban strategy (albeit without a specific plan), in terms of political control, and in terms of sustained public investment. It is therefore ironic that the implementation of the Government's support for a Local Enterprise Company for Glasgow may lead to a break in the consistency of policy and program delivery that has been of considerable benefit to the city.

At stake for Glasgow is the extent to which the private sector will be capable of sustaining the modest improvements of the past 20 years. For the central lesson of urban change in Glasgow has been one of private investment following a clear public lead. Remove that stimulus, and the momentum of change may be lost. Sustain the level of funding, control the ambitions of business leaders, and link central city expansion with the problems of periphery, and Glasgow may indeed flourish in the twenty-first century.

REFERENCES

Barnekov, T., Boyle, R., & Rich, D. (1989). *Privatism and urban policy in Britain and the United States*. Oxford: Oxford University Press.

Boyle, R. (1989, March/April). Partnership in practice: An assessment of public-private collaboration in urban regeneration—a case study of Glasgow Action. *Local Government Studies*, 17-28.

Boyle, R. (1988). Glasgow's growth pains. *New Society, 83* (1306), 15-17.

Bruce, R. (1945). *First planning report to the Highways and Planning Committee of the corporation of Glasgow*. Glasgow.

Bryson, B. (1989, July 9). Glasgow isn't Paris but . . . . *New York Times Magazine*.

Centre for Environmental Studies (CES Ltd). (1985). *Outer estates in Britain*. London: Centre for Environmental Studies.

Checkland, S. G. (1981). *The upas tree: Glasgow 1875-1975 . . . and after 1975-1980*. Glasgow: Glasgow University Press.

Cunnison, J., & Gilfillan, J. B. S. (Eds.). (1958). *The third statistical account of Scotland: Glasgow*. Glasgow: Collins.

Donnison, D. D., & Middleton, A. (Eds.). *Regenerating the inner city: Glasgow's experience*. London: Routledge and Kegan Paul.

Edward Erdman. (1990). *Property investment in Scotland*. Glasgow/London: Erdman Research.

Gibb, A. (1983). *Glasgow: The making of a city*. London: Croom Helm.

Glasgow District Council. (1990, Winter). *Glasgow economic monitor*.

Glasgow District Council. (1989a, Summer). *Glasgow economic monitor*.

Glasgow District Council. (1989b). *Glasgow factsheets 1989*.

Glasgow District Council. (1987a, July). *Unemployment within Glasgow by local area*.

Glasgow District Council, Department of Housing. (1987b). *House conditions survey 1985*.

Glasgow District Council, Department of Housing. (1986a). *The Grieve report—an inquiry into housing in Glasgow*.

Glasgow District Council, Department of Planning. (1986b). *Central area local plan—draft written statement*.

Gretton, J. (1972, October 19). Glasgow: A city in collapse. *New Society*, pp. 138-142.

Industry Department for Scotland. (1988). *Scottish enterprise*, Cm 534. Edinburgh: HMSO.

Jephcott, P. (1971). *Homes in high flats*. University of Glasgow Social and Economic Studies. Edinburgh: Oliver and Boyd.

Keating, M. (1988). *The city that refused to die—Glasgow: The politics of urban regeneration*. Aberdeen: Aberdeen University Press.

Keating, M., & Boyle, R. (1986). *Remaking urban Scotland.* Edinburgh: Edinburgh University Press.

Keating, M., & Midwinter, A. (1983). *The government of Scotland.* Edinburgh: Mainstream.

Kellas, J. G. (1975). *The Scottish political system* (2nd ed.). Cambridge: Cambridge University Press.

Lever, W., & Mather, F. (1986). The changing structure of business and employment in the conurbation. In W. Lever & C. Moore (Eds.). *The city in transition. Public policies and agencies for the economic regeneration of Clydeside.* Oxford: Clarendon.

Local Enterprise Company Unit. (1989). *Toward Scottish enterprise: The handbook.* Edinburgh: Industry Department for Scotland.

McArthur, A. (1987). Jobs and incomes. In D. Donnison & A. Middleton (Eds.). *Regenerating the inner city: Glasgow's experience* (pp. 72-92). London: Routledge and Kegan Paul.

McKie, R. (1989, December 31). Glasgow throws off its hard image. *Observer.*

McKinsey and Co. (1984). *Glasgow's service industries—current performance.* A Report to the SDA. Glasgow: SDA.

McLay, F. (Ed.). (1988). *Workers city: The real Glasgow stands up.* Glasgow: Clydeside Press.

Maclennan, D. (1987). Housing reinvestment and neighbourhood revitalisation: Economic perspectives. In B. Robson (Ed.). *Managing the city.* London: Croom Helm.

Maclennan, D. (1983, November/December). Housing rehabilitation in Glasgow. Progress and impacts since 1974. *Housing Review.*

Massie, A. (1989). *Glasgow: Portraits of a city.* London: Barrie and Jenkins.

Mulgan, G. (1989). New times: A tale of new cities. *Marxism Today, 33* (3), 18-25.

Oakley, C. A. (1975). *The second city* (3rd ed.). Glasgow: Blackie.

Richard Ellis. (1986). *Central Glasgow: A business and office profile.*

Scottish Office. (1988). *New life for urban Scotland.* Edinburgh: Scottish Office.

Smout, T. C. (1986). *A Century of the Scottish people, 1830-1950.* London: Collins.

Strathclyde Regional Council (no date). *Social strategy for the eighties.* Glasgow: Strathclyde Regional Council.

Wannop, U. A. (1986a). Regional fulfillment: Planning into administration in the Clyde Valley 1944-1984. *Planning Perspectives, 1*, 207-229.

Wannop, U. A. (1986b). Glasgow/Clydeside: A century of metropolitan evolution. In G. Gordon (Ed.). *Regional cities in the UK, 1890-1980* (pp. 83-98). London: Harper and Row.

7

Regeneration in Sheffield: From Radical Intervention to Partnership

PAUL LAWLESS

INTRODUCTION

The evolution of urban governance in Sheffield during the 1980s was remarkable. In the first half of the decade the city, with a small group of other British local authorities, developed a radical critique of and alternative to the economic liberalism of Mrs. Thatcher's governments. The city became a laboratory for radical innovation and experiments in urban regeneration. Its activities attracted enormous academic and political attention. For many on the British left, Sheffield was a beacon of hope, an exposer of the moral decrepitude of rampant capitalism and a repository of alternative socialist ideas. Yet by the late 1980s this idea had virtually disappeared. The language of partnership, image, and market investment dominated local debate. Ironically, Conservative central government, which had identified the city as a metaphor of the ills of municipal radicalism, now saw Sheffield as a classic example of how public-private sector alliances might work in British cities. This chapter attempts to chronicle and assess this transformation. To do this, it is necessary to identify some of the major contextual forces that have changed urban governance in the city in recent decades.

THE CONTEXT OF URBAN REGENERATION

A number of forces have, to varying degrees, consistently affected policies designed to regenerate Sheffield. The effect of these forces, as later discussions will make clear, has varied over time. Broadly, it can be argued that the kinds of forces operating on the city in the mid-1970s, before severe economic recession occurred, were markedly different from those found in the late 1980s. Any attempts to assess the scale and direction of recent regeneration policies has to be placed in an appropriate contextual framework. Two major contextual forces merit comment: change in the local economy and dominant political realities.

ECONOMIC CHANGE

For much of the twentieth century Sheffield was a manufacturing city. For more than a century prior to the late 1970s, the economic base of the city was rooted in steel and heavy engineering. The formative processes that encouraged this development have been discussed elsewhere (Foley & Lawless, 1985; Lawless, 1986; Lawless & Ramsden, 1989a). In brief they include the crucial role of a small number of entrepreneurial industrialists in the late nineteenth century establishing vertically integrated companies providing direct downstream outlets for basic steel; the impact of two world wars in enhancing demand; technological innovation; and, in part, investment from and intervention by the state as a result of nationalization in 1967. But whatever the factors responsible for the development of steel and heavy engineering, it is clear that by the 1970s any comparative or historical advantages the city had in the production of steel and heavy engineering goods were fast disappearing. Of a total work force of about 300,000 in 1971, 82,000 people worked in metal goods, engineering and vehicle manufacture, and in other manufacturing industries. By 1986 that figure had fallen to 40,000. In that same period mining employment declined from more than 56,000 to just 15,000, and employment in energy and water supplies fell by half. A recent estimate suggests that by 1996 employment in manufacturing will have fallen to 20% (PACEC, 1989).

Other European and North American cities have endured manufacturing declines on a similar scale. But a number of features make the decline of Sheffield manufacturing particularly interesting. Three issues merit comment: the city's industrial structure, the domination of the public sector, and declining indigenous control of production.

First, it is apparent that the city's industrial structure remains skewed toward sectors such as metal manufacture that are declining nationally. Sheffield has a much lower representation in sectors that have expanded in the country as a whole, notably business services, financial services, and high-tech sectors. This is especially true of private services, which are particularly underrepresented in the city. This partly reflects domination in these financial and allied services by regional competitors, especially Leeds and, to some extent, Manchester. These cities have retained their traditional role as regional centers accommodating a much wider range of public and private-sector services than has Sheffield. For example, Leeds has 50% more workers employed in computer services than has Sheffield; Manchester has 100% more (Henneberry & Lawless, 1989). Sheffield's sole coup, in terms of private services, took place in the mid-1970s when Midland Bank, one of the four major clearing banks in the United Kingdom, transferred many routine functions to the city. By 1988 the bank employed about 2,400 people in Sheffield. But in the city as a whole, relatively little in the way of additional private-sector service employment was evident in the 1980s. It is not, therefore, surprising that professional and managerial jobs declined by 4% in the city between 1980 and 1988, admittedly at a time when semiskilled and skilled manual labor jobs declined much more severely, by 26% and 37%, respectively (PACEC, 1989).

Second, employment in Sheffield remains dominated by the public sector. In 1988 half of the 20 largest employers were in the public sector (Gibbon, 1989). The local authority employed more than 24,000 and the Area Health Authority about half that figure. The Training Agency (previously the Manpower Services Commission), the city's one major success in attracting public-sector organizations from London, employed about 2,000. The largest private-sector employer, British Telecom, employed 2,600. And third, processes of corporate restructuring, especially within steel and heavy engineering, have steadily reduced local control of production, in line with other provincial cities. In 1976, 22 of the 1,000 largest UK firms had their headquarters located in the city; by 1987 that figure had fallen to 13 (Watts, 1989). By 1989 just one of the North of England's largest 100 firms was sited in Sheffield.

Sheffield is the epitome of what has been called a "heavy industry/branch plant labour market" (Cooke, 1986). It has traditionally been dominated by the manufacture of steel and heavy engineering products with a generally skilled and relatively well paid labor force. But during the late 1970s a substantial proportion of the employment disappeared, local control over production diminished, and relatively little growth occurred in those service sectors that were expanding nationally.

POLITICAL RELATIONSHIPS AND REALITIES

Sheffield has been dominated by the Labour Party for many years; apart from two years, the Party has controlled the city since 1926. Typically, two-thirds of the councillors are Labour and five of the six parliamentary constituencies have usually returned Labour members. The sixth, in the western suburbs of the city, has always returned a Conservative. The anomaly reflects the socioeconomic composition of this sector of the city; the western suburbs have one of the highest proportions of professional and managerial heads of household and one of the best-educated electorates of any constituency in the country (Census, 1981).

However, this pattern has not been repeated elsewhere in this notably spatially segregated city. Local and national politics within the city were dominated from the 1920s to the early 1980s by the Labourist tradition (Hampton, 1970). The local economy, based largely on the production of steel and heavy engineering in large plants, provided an environment within which craft organizations and trade unions flourished (Child & Paddon, 1984). Some unions, notably the Amalgamated Engineering Union, played an important role in fostering the Labour Party locally. They sponsored Labour candidates, give financial assistance, and often provided the most active and influential members at Ward and Constituency level.

The Sheffield Labour Party, which the manual unions sustained from the 1920s to the 1970s, introduced a range of municipal socialist policies, primarily in areas of consumption. A major public house building program was started in the 1920s, and Sheffield was active in the 1950s and 1960s in slum redevelopment. Educational services were expanded; in the 1930s, for instance, six new secondary schools were built. Public resources were invested in libraries and art galleries, and the council undertook a number of environmental improvements (Seyd, 1988). In 1957, for instance, smoke control legislation was implemented within the city. In 1974, when the metropolitan counties were created, the new Labour-controlled South Yorkshire County Council introduced a cheap transport system, which, by holding fares steady throughout the 1970s, meant real costs for users diminished sharply. Uniquely among British cities, the use of public transport services increased sharply in the 1970s.

For much of the post-1945 period and until the early 1980s, relationships between the city council and the local business community were relatively uncontentious. The council dominated local politics, and one result of its municipal socialism was that local property taxes were some of the highest in the country. The council's policy of slum redevelopment also removed many older, smaller industrial concerns, many of whom were in the cutlery trade, which declined rapidly in the city between 1960 and 1980. But in other respects, relationships between the authority and the local business community proved mutually beneficial. The council provided land for expansion of larger steel and engineering plants. Public and private sectors cooperated in the redevelopment of the city's retail core. And a limited form of boosterism characterized the council's attitude toward new retail and commercial development; if institutions wished to invest in property, the authority generally granted planning permission. However, Sheffield never explicitly attempted to attract mobile investment.

Political conflict between the local business community and the authority was minimized by the fact that council policy focused primarily on consumption. The major policy debates were concerned with housing, education, social services, and transport. In line with other local authorities, Sheffield viewed production as beyond its remit. The council had a valid role to play in zoning, acquiring, and servicing industrial land, but little else. Policies for production lay with the market or with national government.

The city's policies of municipal socialism, from the 1920s to the late 1970s, were electorally popular. The proportion of the electorate voting Labour at national elections fell steadily from a peak of 57% in 1950 to 35% in 1983. But this decline did not affect the parliamentary representatives, who remained overwhelmingly Labour. Labour usually returned a large majority of councillors to the local authority. However, in the late 1960s Labour councils throughout Britain suffered from the policies introduced by Harold Wilson's Labour government. Cuts in public expenditure and a reduction in real wages were seen by many Labour supporters as unacceptable. In Sheffield one result was that the Conservatives briefly held power in 1968-1969. But the longer-term implications were more profound. As in a number of other cities, a "new urban left" (Gyford, 1985) emerged in Sheffield. Many Labour political activists came from local working-class backgrounds, but most had benefited from higher education and worked in professional, not manual, occupations. This group gained power in the local Labour Party in 1980 and was committed to enhanced public-sector intervention and expenditure on consumption and, crucially, production too. This commitment to intervention in production catapulted the city into a decade of extraordinary political change.

THE ERA OF RADICAL INTERVENTION: 1980-1985

The emergence of a radical group in the local Labour Party, which had an influential role in policy in the first half of the 1980s, coincided with the onset of severe economic recession. In 1979 unemployment in the city stood at less than 5%, slightly lower than the national average. But four years later, the official unemployment rates had almost trebled to 14%, marginally above the national average. During much of the late 1970s and early 1980s more than 1,000 manufacturing jobs were being lost each month in Sheffield.

This scale of industrial decline was a necessary but by no means sufficient cause for the emergence of radical policies in the early 1980s. Other Labour cities enduring equally disastrous economic decline retained orthodox policies designed to attract market investment. A small number of authorities, however, including the Greater London Council, West Midlands, Sheffield, and, to a lesser degree, Leeds and Manchester, followed very different policies. The explanation of why this small group of councils elected to adopt interventionist economic policies remains subject to considerable debate (Blunkett & Jackson, 1987; Goodwin, 1989). The failure of national Labour governments to reduce inequality and implement economic and social reform during the late 1970s clearly encouraged the radicalization of some local Labour parties. Their view was that, if the change could not be implemented by a national Labour government, it might be done locally.

The scale of economic retrenchment introduced by Mrs. Thatcher between 1979 and 1981 was also a major feature. High interest rates, a substantial reduction in infrastructural investment, and an overvalued pound were a lethal concoction for many traditional manufacturing sectors in the big cities. In Sheffield the drive toward local intervention was encouraged by the direct election of trade union representatives onto the District Labour Party, which regulated city council policy. They made the realities of industrial decline and massive job contraction very apparent to the local Party.

Whatever the reasons for this radical drift in many Labour councils, the early 1980s witnessed a dramatic growth in analyses of, and policy intervention within, local economies. Both sides of this equation, analysis and prescription, were well developed in Sheffield.

THE RADICAL PROBLEM DEFINED

From the early 1980s the city council and its Trade Union allies produced a series of sectoral analyses exploring the root causes of economic decline (Sheffield City Council, 1982, 1984, 1987a; Sheffield Trades Council, 1982). Some of this work, notably the material on steel in 1984 and on special steels in 1987, was of a sophisticated and academic nature. The details of these publications need not detain us. But the central message was very clear: Steel and heavy engineering in Sheffield were subject to a range of international forces, over which the local community had no control. Decisions about production, investment, employment, and marketing were very largely undertaken by public- and private-sector organizations located outside the city.

The analysis was, of course, not original. The National Community Development Projects had said similar things in the mid-1970s (National Community Development Project, 1977). However, the local authority was particularly impressive in teasing out the impact of a range of intermeshing processes on the production of steel and heavy engineering goods. Some of these processes concerned the international trade in steel. For example, newly developing countries were undercutting basic steel production costs throughout Europe and North America. Quotas imported by the European Commission affected the production of some products. Investment patterns were crucial to understanding steel decline too. The nationalized sector, the British Steel Corporation, had overinvested in basic steel production to the detriment of higher value-added products. And investment in the private sector had done little to boost productivity and profitability in special and alloy steels. But the main culprit, in the eyes of Sheffield City Council, was central government. Its policies in the late 1970s and early 1980s had proved disastrous for the city. Macroeconomic policy had induced a severe recession in steel-using sectors; public investment in the nationalized sector had fallen sharply; efforts to privatize parts of the steel industry, through the so-called "Phoenix" mergers, had culminated in massive rationalizations, loss of output and employment, and a virtual handing over of public assets to the private sector.

The interpretation did not pass unchallenged. Other commentators argued that the rapid contraction of employment and output in steel was an inevitable consequence of poor management, overmanning, restrictive practices, and unwise investment (Aylen, 1984). But in Sheffield the high ground of debate was secured by the council. Little to contradict its perspective was heard from the local business community, which, in any case, was split in its attitudes. Many business leaders had consistently argued for a shakeout in industry; however, few welcomed the scale of retrenchment that occurred between 1979 and 1982. In any case, the larger redundancies occurring within the city were largely not effected by local managers. Those decisions were made elsewhere. This made the authority's propaganda role easier: Contractions in steel and heavy engineering could be firmly laid at the feet of national government and external management.

THE RADICAL ERA: POLICY INNOVATION AND DEVELOPMENT

Economic decline and employment contraction were central to the political debate in Sheffield in the early 1980s. Traditionally, consumption issues such as housing, transport, and education had framed much political activity. But the scale of retrenchment, in a city that had prospered throughout the long postwar boom, inevitably redirected attention toward the economy and jobs. Much of the analysis that emerged in the late 1970s and early 1980s focused on the local economy. The local economy also proved to be the major focus for policy development.

In 1981 the city created an Employment Department, whose first Director was a radical political economist. The Department was given a very wide brief (Sheffield City Council, 1982): to coordinate the council's activities in order to prevent further job loss; to alleviate the worst effects of unemployment; to stimulate new investment; to create new types of jobs; and to explore new forms of industrial democracy. It was given a limited initial budget of £2.5 million per annum.

In its early development the Department concentrated on a number of key program areas. For example, a Research and Resources Unit was established to develop early warning signs of job losses and assist trade unions and other groups in resisting job losses. Equal opportunities for women and other groups were pursued. Efforts were made to enhance local training and technological innovation. The Department also elected to provide financial assistance to enterprises and seek out investment opportunities in the local economy, where possible with appropriate financial institutions.

It is important to stress that the creation of an Employment Department was as much a political initiative as an economic one. To its first Director, John Benington, the primary aim of the Department was "to liberate the resources of the local state and put them at the service of the working class movement, the women's movement and community based movements" (*Critical Social Policy*, 1984, p. 83). The Department was helping to restructure the economy to the benefit of labor, not capital. It was to open opportunities for the production of socially useful goods in democratically controlled organizations. Although there was a degree of realism about the real scale of intervention possible for any local authority, there was, nevertheless, a sense that what was happening in Sheffield would initiate discussion in the Labour Party of the role of the state in both local and national economics.

An evaluation of the Employment Department's activities in the first half of the 1980s must make a distinction between broader strategic issues and program implementation. In terms of the latter, many projects initiated by the Council proved effective and innovative (Blunkett & Jackson, 1987; Cochrane, 1988; Sheffield City Council, 1982). A number of cooperatives were assisted in their formation by the authority. Training for those within the council was enhanced, and a range of new socially useful products, including equipment for the disabled, an advanced humidifier, and software for the blind, were developed in a product development unit supported by the Council.

However, these limited benefits have to be set against some real difficulties (Abel, 1985; Cochrane, 1988; Goodwin, 1989). There were tensions between the Department and other more "pragmatic" sections within the authority. Not all councillors were eager to embrace the "top-down" approach adopted by some officers in the Department; the city had always stressed that economic development should emerge from an active community base (Blunkett & Green, 1983).

Crucially, too, the strategy did not work. It could not work. The resources available to the Department, £18 million in its first seven years of existence, were minute in the context of investment patterns in steel and heavy engineering. Other councils in the country managed to acquire larger resources by using their own pension funds to invest in equity and by creating independent enterprise boards. Sheffield's pension fund was not controlled by the city, but by the Metropolitan County, and it elected not to create an enterprise board. These constraints undoubtedly reduced already severely circumscribed powers of intervention. But although the reality of intervention was a long way divorced from the rhetoric, the latter had one obvious effect: It dampened down any enthusiasm the market had for investing in the city. Hardly any retail, commercial, or industrial development took place in the city in the first half of the 1980s. The Sheffield Department of Employment's experiment in radical local intervention in the early 1980s attracted remarkable academic and institutional interest; the beneficial effects for communities and individuals are harder to identify.

THE EMERGING PARTNERSHIP: 1986-1989

In 1983 the local Labour Party argued that "we now need to claim for a locally elected authority like Sheffield the right for greater community and workers control and influence over employment and the local economy" (Employment Manifesto Working Group, p. 1). Yet just five years later the leader of the council could argue that the public and private sectors working together in Partnership had been able to break down barriers and bring about cooperation in a range of projects likely to provide jobs for the city (Betts, 1988). Somewhere in the mid-1980s a substantial change in attitudes took place within the city, as Table 7.1 reveals. The substantial changes of direction in the mid-1980s raise three crucial, related issues: the root causes of the emergence of Partnership; its structure, evolution and policy development; and its impact.

WHY PARTNERSHIP?

In examining the transformation in attitudes toward development in the mid-1980s, a distinction can be made between immediate local effects and broader national impulses. At the local level a number of issues emerged in the middle years of the decade. The previous Leader of the council, Blunkett,

TABLE 7.1
Strategies for Economic Regeneration

Strategy	*Phasing*	*Key Objectives*	*Policy Instruments*	*Funding*	*Agencies for Renewal*
"Labourist Tradition"	To 1979/80	Enhanced Investment in collective Consumption	New Housing; Educational Investment; cheap transport; zoning and servicing land.	Local and Central Government; Nationalised industries	Local Government in aspects of Consumption; national Government, nationalised industries and private sector in production.
"Radical" Era	1980-85/86	Public Sector Intervention in Local Economy; Job creation; Proselytizing Function	Employment Department; Financial Equity in Firms; Co-ops; Equal opportunities; Product Development, etc.	Local Authority	Local Authority
"Partnership"	1985/86-1989	Collaborating Programme of Economic Development; Improving image of city; better "Business Environment"	Flagship Projects in Retailing, Leisure, Commercial development, etc.	Largely Private; "Pump-priming" Local and central Government Expenditure	Local Authority, Private Sector, through Sheffield Economic Regeneration Committee (SERC); Sheffield Development Corporation

was elected to Parliament and was replaced by Clive Betts. At the time this was not regarded as significant. In retrospect Betts proved to be a pragmatic leader in charge of an authority that increasingly adopted collaborative relationships with business. Other changes in key personnel occurred. The first head of the Employment Department left and was replaced by a less ideological figure. In 1986, after the abolition of the Metropolitan County of South Yorkshire, the city assimilated a number of officers from the defunct authority who were more development oriented than their city counterparts. A number of these officers later attained powerful positions within the council. Also, these changes in personnel took place when the intellectual sophistication but practical irrelevance of radical municipal intervention had become very apparent.

Although local forces were important in focusing attention on the merits and defects of radicalism, these debates were conducted within a wider context of national economic and political realities. Sheffield's experiment in radical local intervention, as with other authorities, especially the Greater London Council, was intended to influence the wider debate in the National Labour Party. These local experiments were assimilated nationally into the Alternative Economic Strategy, which was rejected by the national electorate in 1983. For Labour in Sheffield this defeat was very significant. The election heralded a period of Conservative hegemony, and the prospect of increased central control of local government.

The city's main political allies also disappeared in 1986 with the abolition of the metropolitan counties and the Greater London Council. From the mid-1980s onwards, the National Labour Party was shedding its more radical policies in an attempt to regain the middle ground of British politics. If Sheffield perpetuated a radical program of local economic intervention, it would travel a lonely road. This became more apparent near the end of the 1980s as central government reoriented its urban program toward the private sector so that in 1988, in its major policy statement "Action for Cities," there was not one single mention of local government (HMSO, 1988). Urban policy would increasingly be devised by, be implemented by, and, in turn, benefit the market.

PARTNERSHIP: EVOLUTION, STRUCTURE, AND POLICY DEVELOPMENT

After 1985 Sheffield City Council and the business community began a process of reconciliation. Much of the initial discussions were informal. On the local authority's side, the personnel involved were the leader of the Council, several leading politicians, and heads of some key departments. Much of the drive for partnership in the private sector came from the

Chamber of Commerce. An effective Chief Executive and a number of active presidents provided the focus for most private-sector activity. As with the city council, few individuals were involved. Fewer than 10 key figures from the private sector initiated and sustained most partnership programs. A large proportion of these were male, middle-aged, chairmen of local companies.

These informal relationships between the city and the Chamber of Commerce led eventually to the creation of the Sheffield Economic Regeneration Committee (SERC) in 1987. This consists of representatives from a range of local interests: the authority, private sector, trade unions, community groups, higher education, and the Sheffield Development Corporation. The last of these institutions was established by central government in 1988, with an initial budget of £50 million, to regenerate the city's traditional industrial area, the Lower Don Valley. However, although SERC embraced a range of local constituencies, the driving force for much that has happened has been the local authority and the private sector. Promotional material features the two organizations; SERC is serviced by the council; conferences are addressed by leaders from both organizations. Other sectors and institutions are much less prominent in SERC and in the larger Partnership.

The emergence of SERC in Sheffield can be placed in the wider debate about the concept of "growth coalitions," an American notion that has recently entered the literature in the UK (Cooke, 1988; Lloyd & Newlands, 1988). The Sheffield case study illuminates three key issues about growth coalitions in Britain. First, in Sheffield, the local authority has been the key driving force behind the growth initiative. In other cities local authorities have not always been so active. Second, in Sheffield the range of interests involved in the Partnership is not great. Constituencies such as central government, the media, and major multinationals, which have played prominent roles in coalitions in other British cities, are not active in Sheffield. The Partnership initiative, reflecting the political and economic structure of the city itself, remains itself parochial. Third, property interests, which in academic literature are identified as a core group in growth coalitions (Molotch, 1976), are not especially significant. Much land in areas likely to be subject to redevelopment is owned by the public sector.

In common with other coalitions, SERC has identified a series of predictable market-oriented issues as constituting the core economic problem of the city. Questions of image, the attractiveness or otherwise of Sheffield to the market, and the need to provide the right environment for private investment to occur have dominated SERC's public announcements. For example, a previous president of the Chamber of Commerce has argued through SERC's promotional material that "Sheffield is a good place in which to make a living, a good place to invest" (Sheffield Partnerships Ltd. and SERC, 1988).

In attempting to overcome the poor market perception of Sheffield, SERC has acted as a catalyst and coordinator in a wide program of activities. But this program does not constitute a strategy. In some cities a series of guiding principles has governed partnership (Boyle, 1989). In Sheffield this has not occurred. A series of themes has been identified by SERC as constituting a vision for the city (Sheffield Partnerships Ltd. and SERC, 1988); but several of them appear overblown or merely a rationalization of past events. The argument, for example, that the city might become a world leader in research technology or a new decision center for Britain appears optimistic. The city has only recently gained a science park, and national and even regional media are not represented.

Nevertheless, despite the diffuse and inchoate nature of SERC's vision, the city's Partnership has presided over a series of initiatives. Mainly, SERC itself has not implemented projects; indeed, it has no specific budget. It operates as a catalyst, a coordinator and a broker, developing schemes by integrating agencies and teasing out resources from a wide range of existing budgets in the public and private sectors. Virtually all of the projects in the field are economic and/or property development. Seven main program areas can be identified: promotional activities, retail and commercial development, infrastructural provision, sports, culture, education and training, and housing.

On the promotional side SERC has held conferences, undertaken visits to London, attempted to attract inward investment by targeting London-based financial institutions. Partly as a result of this activity, the city has seen an upsurge in property development in the past three years. Between 1986 and late 1988 more than 100 £1 million-plus schemes had been granted planning permission (Sheffield City Council, 1988a). Many of these were relatively small scale but one, Meadowhall, sited on a previous steel works, is a £200 million retail and leisure complex. Most of this investment has emerged from financial institutions, although a small number of projects have attracted Urban Development Grants (now City Grants).

Probably the most surprising of the new developments, however, is the decision by the International Federation of University Sports (FISU) to hold the 1991 World Student Games in Sheffield. The WSG, although virtually unknown in Britain, is the second-largest athletic and cultural gathering after the Olympic Games. Considerable doubt has been expressed about the viability of the scheme (Rodda, 1988). However, by summer 1989 capital investment in excess of £100 million had been raised for the Games. About one-fifth of this will come directly from the private sector, notably for the construction and the operation of an indoor arena. More than £78 million is being borrowed by the city, however, from a consortium of banks. This raises longer-term questions about capital repayment and revenue charges on facilities after the Games.

During the Games a cultural festival will be held to reinforce the city's determination to expand in this area. The city has already developed a cultural industries quarter, where almost £3 million from the Urban Programme has been used to convert older industrial and commercial buildings into municipal studios, an audiovisual center, photographic studios, and other cultural facilities (Planning Exchange, 1989).

Finally, two other areas ought to be mentioned. A number of education and training initiatives have been created under the umbrella of the Sheffield Education/Business Partnership. These include compacts between schools and local employers, industrial secondments for teachers, and employer links with schools. And the city has made an agreement with the United Kingdom Housing Trust to develop 4,000 properties for rent and sale, funded by bank loans, mortgages, housing association grants, and building society mortgages.

SHEFFIELD PARTNERSHIP: AN EVALUATION

Evaluating the Sheffield Partnership raises a number of methodological problems. For example, collaboration within the city has been, as with other similar initiatives, vague about objectives and activities. It is trying to enhance market perceptions of, and willingness to invest within, the city. But it has not defined either how this will occur or how its success might be assessed. Moreover, the movement toward Partnership has been diffuse: SERC helps coordinate and further at least 21 separate projects. Some of these may well be successful in their own terms but do little to foster the wider goal of public-private sector collaboration. Finally, there is a crucial issue of additionality. How much of what is happening in Sheffield would have occurred irrespective of Partnership?

Despite these caveats, some attempt at evaluation should be made. Sheffield's Partnership has undoubtedly made some advances. Relations between the city and the private sector have improved substantially since the early 1980s. Relations between the city and central government have altered, too. In recent years the city has received support from English Estates, central government's industrial development agency, for the construction of the science park; a number of Urban Development Grant applications have been approved; an Urban Development Corporation has been established, which, although resisted by the council initially, was eventually accepted and a formal agreement signed between the city and the Corporation. And promotional materials featuring the city and local Chamber have received formal blessing from the Prime Minister, Mrs. Thatcher (1988). It is hard to imagine many of these developments occurring in the early 1980s, when the city council was committed to the "Socialist Republic of South Yorkshire."

In light of central government's redefined attitudes toward Sheffield, it is not surprising that the market (Employment Times, 1989) takes a more favorable view of the city than in the early 1980s: Sheffield's image has improved within the market, which has helped encourage the scale of retail and commercial developments in the latter years of the decade. However, since virtually no commercial development occurred between 1980 and 1985, local property agents had identified a shortfall of almost 1 million sq. ft. in the city (Employment Times, 1989). Speculative developments are far more attractive in such an environment.

In contrast to the benefits Partnership has brought to the city, a longer-term assessment may prove more critical. Five themes merit consideration here: efficiency, equity, the robustness of Partnership, its impact, and its diversionary nature.

One of the characteristics of urban regeneration that is based on market precepts is the ad hoc, pragmatic nature of much development. In national terms, this is especially true of London's Docklands (Lawless, 1989), but aspects of this problem can also be found in Sheffield. A great deal that has happened in the city has been the implementation of major flagship developments. While they are generally perceived as helping the city, they do raise a number of considerations. Longer-term land use planning becomes virtually impossible (Lawless & Ramsden, forthcoming). Land is too rapidly allocated for specific uses, which may have limited longer-term benefits for the city as a whole. In addition, major market-sector developments may require extra public-sector investment beyond that which would have otherwise occurred.

The full longer-term financial implications of particular projects may be unknown (Stone, 1987). This is likely to prove the case, for example, with the World Student Games. Their "success" may well require additional public investment. On the capital side, it appears probable that a new Super-Tram will be constructed through the city to the main arenas. On the revenue side, resources may well be required for both expenditure in the run up to 1991, and to repay the interest on borrowed capital after the Games take place. The raising of this expenditure will fall disproportionately on less-affluent groups in the city.

Many commentators have raised equity issues in relation to urban regeneration led by public/private alliances (Boyle, 1989; Docklands Forum, 1989; Edwards, 1989). Certainly in Sheffield recent developments have had considerable distributional effects. Particular groups have benefited more than others during Partnership. Local industry has done well; profitability, output, and investment in steel and heavy engineering rose sharply from 1985 to 1987 (SERC, 1987). The owners of development land saw a quadrupling in values in 18 months during 1988-1989 (Employment Times, 1989). Key business personnel now dominate SERC and the board of the Sheffield Development Corporation, and will play a crucial role in the new Training and Enterprise Council.

Prospects for other sectors in society remain bleak (Employment Service, 1989; Sheffield City Council, 1988b). While official statistics show a fall in unemployment from a peak of 48,000 or 17% in 1986 to 32,000 or 11% in early 1989, that still represents an unemployment total 250% higher than the level in 1979. Moreover, in some parliamentary constituencies, unemployment exceeds 20%, and equivalent figures for Black and Asian people stand at almost 50%. Long-term and youth unemployment were falling only slowly at the end of the decade, and increasing as a proportion of total unemployment.

With an anticipated continued decline in skilled and semiskilled manual jobs, and with Partnership largely ignoring questions of collective consumption, the Partnership has proved largely irrelevant for many of those previously working in, or who would otherwise have found employment within, steel and heavy engineering. The Leader of the Council has repeatedly addressed this problem (Betts, 1988). Indeed, some programs designed to allocate a proportion of construction jobs to local people have been devised. But the overall message from SERC, the Development Corporation, and indeed the city council relegates distributional questions to a secondary status.

A third issue about Partnership is the question of its relative robustness. The partners came together in the mid-1980s, when the local economy was subject to the most severe recession since the late 1920s, and the alliance has apparently lasted well. But divisions and conflicts remain. At one level there are disagreements between partners. For example, despite entering into formal agreement with the Development Corporation, the City Council has not always been awarded agency contracts in areas such as land use planning. Central government has not always agreed with SERC's recommendations. SERC's first act was to commission a consultancy of the Lower Don Valley, which recommended the creation of a separate development company to oversee the strategic regeneration of the area. But despite putting resources into the consultancy, central government ignored the recommendation and instead imposed an Urban Development Corporation on the area.

Apart from divisions between partners, there have been splits within specific sectors. The business community, a somewhat heterogeneous category, shows marked divisions. The Development Corporation is proposing a major road scheme to attract commercial development, but has received considerable criticism from smaller industrial concerns, some of which are likely to be compulsorily acquired. Ultimately, urban regeneration is not an apolitical, neutral, or technical activity. Decisions will affect different sectors in contrasting ways. The economic expansion in the city in the second half of the 1980s helped smooth over divisions. But any downturn in the local economy in the early 1990s is likely to open political debate and conflict.

A fourth issue of Partnership is the question of additionality: To what extent has Partnership been responsible for the scale of development in the city in the late 1980s? It has probably helped in attracting inward investment, which has largely been directed toward property development. But many of the projects that SERC coordinates might have been implemented anyway. Several of them, such as the Development Corporation, the Science Park, the Sheffield Enterprise Agency, and the World Student Games, either predate SERC, or depend for their implementation on specific budgets and agencies, or involve minimal contribution from SERC. Indeed it might be argued that the upturn in the local economy in 1986-1987 gave rise to conditions within which a collaborative venture like SERC might thrive. SERC was perhaps as much an effect as a cause of the city's development boom.

Finally, there is the issue of the diversionary nature of the Partnership exercise. The city has received wider, more favorable media coverage in the past few years than in the early 1980s. Central government has also praised the degree of public-private sector collaboration in Sheffield. But such reactions avoid other crucial issues for the inhabitants of the city. The council estimated in 1987, for instance, that £650 million was required to modernize its 89,000 dwellings, and at least £70 million to refurbish educational facilities (Sheffield City Council, 1987b). Yet throughout the 1980s the council's ability to undertake capital expenditure was severely eroded by central government. On the revenue side, central government's support for local government services has fallen in real terms. There was, for instance, virtually no increase in the government grant between 1988-1989 and 1989-1990 despite inflation, which imposed cuts in excess of £12 million in revenue expenditure by the authority. While this loss amounts to only 2% of the total city expenditure, it is more than double the annual Urban Programme allocation granted to the city. This occurred at a time when demand for welfare and social services, notably for the old and infirm, is increasing sharply, and in the inner-city area of 200,000, at least 20% of the economically active are unemployed. Private-public sector partnerships neatly fit that series of urban initiatives effected in the United Kingdom for more that 20 years (Lawless, 1986b): They address certain issues, they marginalize others.

REFERENCES

Abel, P. (1985). Shelling out the cash in Sheffield. *Town and Country Planning, 54* (6), 194-195.

A parable of how things might be done differently. (1984). *Critical Social Policy, 9*, 69-87.

Aylen, J. (1984). Prospects for steel. *Lloyds Bank Review,* (152), 15-23.

Betts, C. (1988, May 23). Building a new future. *The Sheffield Star; Sheffield Comes to London Supplement.*

Blunkett, D., & Green, C. (1983). Building from the bottom: The Sheffield experience. *Fabian Tract 491.* Fabian Society.

Blunkett, D., & Jackson, K. (1987). *Democracy in crisis: The town halls respond.* London: Hogarth Press.

Boyle, R. (1989). Partnership in practice; an assessment of public-private collaboration in urban regeneration—a case study of Glasgow Action. *Local Government Studies, 15* (2), 17-28.

Census. (1981). Office of Population Censuses and Surveys. County report, South Yorkshire. London: HMSO.

Child, D., & Paddon, M. (1984). Sheffield: Steelyard blues. *Marxism Today, 28* (7), 18-22.

Cochrane, A. (1988). In and against the market? The development of socialist economic strategies in Britain, 1981-1986. *Policy and Politics, 16* (3), 159-168.

Cooke, P. (1986). Global restructuring, industrial change and local adjustment. In P. Cooke (Ed.). *Global restructuring, local response* (pp. 1-24). Economic and Social Research Council.

Cooke, P. (1988). Municipal enterprise, growth coalitions and social justice. *Local Economy, 3*, 191-199.

Docklands Forum. (1989). *Does the community benefit? What can the private sector offer?* London: Docklands Forum.

Edwards, J. (1989). Positive discrimination as a strategy against exclusion; the case study of the inner cities. *Policy and Politics, 17* (1), 11-24.

Employment Manifesto Working Group. (1983). Introduction; Sheffield City Council Labour Group.

Employment Service. (1989). *Labour market information.* Sheffield/Rotherham Area.

Employment Times. (1989, March 10). South Yorkshire: An Estates Times Survey.

Foley, P., & Lawless, P. (1985). *Sheffield: Economic change 1971 to 1985 and policies to reduce decline.* School of Urban and Regional Studies, Sheffield City Polytechnic.

Gibbon, P. (1989). *Capital restructuring in Sheffield.* Department of Applied Social Studies, Sheffield City Polytechnic.

Goodwin, M. (1989). The politics of locality. In A. Cochrane & J. Anderson (Eds.). *Restructuring Britain; politics in transition* (pp. 141-171). London: Sage.

Gyford, J. (1985). *The politics of local socialism.* London: Allen and Unwin.

Hampton, W. (1970). *Democracy and community.* Oxford: Oxford University Press.

Henneberry, J., & Lawless, P. (1989). *Advanced producer services in areas of manufacturing decline; the computer services sector in Sheffield* (Working paper No. 16). School of Urban and Regional Studies, Sheffield City Polytechnic.

HMSO. (1988). *Action for cities.* London: HMSO.

Lawless, P. (1986a). *Severe economic recession and local government intervention: A case study of Sheffield.* School of Urban and Regional Studies, Sheffield City Polytechnic.

Lawless, P. (1986b). *The evolution of spatial policy.* London: Pion.

Lawless, P. (1989). *Britain's inner cities* (2nd ed.). London: Paul Chapman Publishing.

Lawless, P., & Ramsden, P. (1989a). *Sheffield into the 1990's; urban regeneration: The economic context* (Working paper No. 20). School of Urban and Regional Studies, Sheffield City Polytechnic.

Lawless, P., & Ramsden, P. (Forthcoming). *Land use planning and the inner cities; the case of the Lower Don Valley, Sheffield.* Local Government Studies.
Lloyd, M. G., & Newlands, D. A. (1988). The "growth coalition" and urban economic development. *Local Economy, 3* (1), 31-40.
Molotch. H. (1976). The city as a growth machine: Towards a political economy of place. *American Journal of Sociology, 82*, 309-332.
National Community Development Project. (1977). *The costs of industrial change.*
PA Cambridge Economic Consultants (PACEC). (1989) *Sheffield employment study: Economic forecasts for the Sheffield City Study Area.*
Planning Exchange. (1989). *Sheffield development; cultural industries quarter; Urban Development Area initiatives A 36.* Glasgow: Planning Exchange.
Rodda, J. (1988, October 21). A hard sell for Sheffield. *The Guardian.*
Seyd, P. (1988). *Socialist city politics.* Sheffield Department of Politics, Sheffield University.
Sheffield City Council. (1982). *An initial outline.* Employment Department.
Sheffield City Council. (1983). *Review of the work in the first year.* Employment Department.
Sheffield City Council. (1984). *Steel in crisis.*
Sheffield City Council. (1987a). *The uncertain future of special steels.*
Sheffield City Council. (1987b). *Working it out; an outline employment plan for Sheffield.*
Sheffield City Council. (1988a). *Major development proposals in Sheffield; land and planning.*
Sheffield City Council. (1988b). *The Sheffield labour market.* Department of Employment and Economic Development.
Sheffield Economic Regeneration Committee (SERC). (1987). *Lower Don Valley, final report by consultants.*
Sheffield Economic Regeneration Committee (SERC). (Undated). *SERC, Sheffield City Council.*
Sheffield Partnerships Ltd. and SERC. (1988). *Sheffield vision.*
Sheffield Trades Council. (1982). *The second slump.*
Stone, C. N. (1987). Summing up. In C. N. Stone & T. S. Heywood (Eds.). *The politics of urban development* (pp. 269-290). University Press of Kansas.
Thatcher, M. (1988, May 23). A message from No. 10. *The Sheffield Star; Sheffield Comes to London Supplement.*
Watts, H. D. (1989). Non-financial head-offices; a view from the north. In J. Lewis & A. Townsend (Eds.). *The north-south divide* (pp. 157-174). London: Paul Chapman.

8

Montreal: The Struggle to Become a "World City"

JACQUES LÉVEILLÉE
ROBERT K. WHELAN

SETTING THE CONTEXT

In a comparative urban context, Montreal is closest to the large U.S. metropolitan areas classified as "regional nodal diversified service centers" in a study of 140 U.S. metropolitan areas (Noyelle & Stanback, 1983). In particular, Montreal compares in many ways with such Northeastern metropolises as Boston, Philadelphia, and Baltimore (Colcord, 1987; Gappert, 1988; Levine, 1989). All of these cities were industrial and manufacturing centers with economies originally based in their ports. All have faced the transition to a service-based economy.

Demographic changes in the Montreal metropolitan area paralleled those in similar U.S. metropolitan areas in the 1971-1981 decade. While the Montreal metropolitan region did not lose population, its growth rate was much slower than that of other Canadian metropolitan regions. Moreover, there was a definite shift in the distribution of the metropolitan area's 2.8 million population that paralleled changes in U.S. cities. The city of Montreal's population dropped by 200,000 to one million. The population of the rest of the island, that is, the Montreal Urban Community (the inner suburbs in the U.S. context) was the area of dynamic expansion in the 1971-1981 decade.

Between 1971 and 1986 the city of Montreal lost 70,000 manufacturing jobs, or more than 36% of its industrial base. In another parallel to the U.S. experience, manufacturing locations shifted from the central city of Montreal to suburban sites. As noted above, manufacturing jobs first shifted to suburbs on the island, such as Dorval and Pointe Claire. In the 1980s, manufacturing growth has deconcentrated further, to such off-island suburban cities as Laval and Longueil. The absolute and relative decline of manufacturing jobs was especially evident in such traditional industries as leather goods, textiles, hat making, and clothing. Theses industries are all labor-intensive, low-value-added industries that were devastated by foreign competition. Overall, between 1951 and 1986, the proportion of the metropolitan Montreal work force employed in manufacturing declined from 37.6% to 21.2% (Lamonde et al., 1988; Levine, 1981; Nader, 1976). In short, a process of deindustrialization occurred in Montreal. Moreover, this process had spatial implications, as manufacturing deconcentrated along both east-west and north-south axes in relation to the central city (Lamonde et al., 1988).

Higgins (1986) notes that "what has distinguished Montreal from the beginning is the city's role on the international scene." It should be noted that this role was a part of the British Empire. Economic ties with Great Britain, along an east-west Canadian axis, were vital for Montreal. Since World War II, the United States has replaced Britain as a focal point for the Canadian economy. In Montreal's case, the integration of Canada's economy with the U.S. was a double blow. Cities, such as Montreal, whose prominence derived from east-west linkages, were bypassed by cities such as Toronto, which were vital for north-south linkages (Gagnon & Montcalm, 1989). Montreal's traditional international role is best demonstrated by its port activity. In 1929 Montreal was the most important port for export of grains in all of North America. In relation to other Canadian ports, Montreal declined in relative importance on such measures as total traffic, international traffic, bulk merchandise, and general merchandise between 1961-1973. Statistics from the mid-1970s further demonstrate the decline of Montreal as a port in relation to other Canadian cities (Higgins, 1986).

In another economic area, Montreal has suffered in relation to Toronto. In the post-World War II period, there were substantial changes in the continental economy, which undermined Montreal's position and its relationship to the rest of the Canadian economy. The general decline of industry and commerce affected the financial sector, particularly in relation to Toronto. In 1946 the Montreal Stock Exchange had 62.8% of the value of transactions of the Toronto Stock Exchange. By 1985 the MSE had only 23.8% of the value of transactions of the TSE (Gagnon & Montcalm, 1989). In 1952 Montreal had 20% more head offices of financial headquarters than Toronto. Another study found that the ratio of company headquarters in Montreal and Toronto changed from 124/100 in 1951 to 62/100 in 1972. Taken together, these trends indicate that Montreal's diminished role in the Canadian and North American economy began long before the *Parti Quebecois* (PQ) came to power in the mid-1970s, long before Quebec nationalism was in vogue, and long before the PQ passed legislation to protect the French language in the province (Gagnon & Montcalm, 1989; Higgins, 1986).

Montreal's unemployment rate has been high in the past two decades. Despite different rates of counting and a higher social net, Montreal has consistently maintained a high unemployment rate. In 1970 the unemployment rate was 10.2% in the Montreal metropolitan area. The rate was as low as 6.1% in 1985, and was 8.8% in 1980. Montreal has been affected by the oil and energy recession of the 1980s, and the unemployment rate has been in double digits since 1982. The highest unemployment rate is the current 14% in early 1990. The unemployment rate has stayed steadily at, or close to, 10%. The question is: How did the city respond to economic transition in the 1970s and 1980s?

MONTREAL: THE INSTITUTIONAL CONTEXT

Montreal is a city of one million inhabitants who coexist with 29 other principalities (700,00 inhabitants) inside a metropolitan structure, the Montreal Urban Community (MUC), whose territory of jurisdiction is the island of Montreal. The Urban Community administers a budget of about $1 billion. The city's budget is about $1.7 billion.

The Community is a federation of cities and is not a metropolitan government on its own. The MUC Council is composed of all the representatives of the city of Montreal (59 persons) and of a representative, generally the mayor, of each of the municipalities on the Island. The decisional process is managed by permitting a veto power for each of the two blocs. The members of the MUC Council carry out a good part of their legislative work in the five permanent Commissions, which work out the politics and supervise the evolution of programs in the principal fields of intervention of the Community. The two principal responsibilities of the MUC are the police and public transportation (about 75% of the budget). The other powers exercised by the MUC are the appraisal of property, economic promotion, planning, and the environment.

The first 12 years of the Community's existence have been lively. The suburban cities on the Island did not succeed in truly accepting the Community. The suspicions of the suburban cities in regard to the city of Montreal were kept alive by an excellent memory of the "imperialistic" strategy that had been pursued by Mayor Drapeau and his team under the slogan "One island, one city." Since 1982, after the last important reform of the MUC, the relations between the two blocs have been much more harmonious. All those involved have become aware that they cannot disassociate themselves from Montreal's destiny in the Canadian, Quebec, and regional network to which it belongs. Rather than trying to leave the body of Montreal in hopes of saving themselves, the Island's suburban cities admit that their fate is intimately tied to the city of Montreal.

Given the city's political arrangements, many would not have projected a positive response to the painful problems of transition from a manufacturing to a service-based economy. The mayor of Montreal for almost 30 years (1954-1957, 1960-1986) was Jean Drapeau. A colorful mayor and a great booster of Montreal as an international city, Drapeau presided over an urban regime that practiced a traditional and personalized politics. Montreal politics were similar to those of urban machines in the eastern and midwestern United States. Moreover, Montreal's most important business leaders in Drapeau's early years were Anglophones. Redevelopment in Montreal was undertaken in the privatist tradition of the U.S. and the UK (Barnekov, Boyle, & Rich, 1989).

The first phase of urban redevelopment in Montreal occurred in the 1950s and 1960s, as commercial and office activity moved from the historical center in "Vieux Montreal" to a new city core centered around Dorchester Boulevard, north of the old central business district. It is important to note this was done without the stimulus of a federal urban renewal program, as existed in the U.S. Furthermore, this was done without any official city plans. The most important of these private redevelopment efforts was Place Ville Marie, which George Nader calls the "single most important development in the history of Montreal's downtown" (Nader, 1976). Place Ville Marie began the development of Montreal's extensive underground system. Other major projects that opened during the 1960s were Place Victoria, Place Bonaventure, and Place du Canada. These were all large, multifunctional commercial and office complexes. Mayor Drapeau worked closely with such developers as the American, William Zeckendorf. Renewal was undertaken by the private sector, with cooperation and accommodation from the city government.

The second phase of redevelopment, in the 1960s and 1970s, revolved around the "grand projects" of Mayor Drapeau. He was a visionary, but he was not a planner. The "grand projects" included: infrastructure improvements (boulevards, highways, the subway); great international events (the 1967 World's Fair, the 1976 Olympic Games); and the development of several *plans d'ensemble* (essentially megaprojects), including the Complex Desjardins, the Complexe Guy Favreau, the Palais des Congres, Place des Arts, Radio-Canada, Radio-Quebec). All of these undertakings were facilitated by Drapeau's use of city powers of zoning and expropriation, with land swaps, and concessions to the developers. All of these projects proceeded and developed in an uncoordinated fashion, without any systematic thinking about their relationship—either to each other or to the urban whole. Planning and zoning proceeded on an ad hoc basis. One study compared Montreal's zoning to Houston's nonzoning system (Sancton, 1983).

Drapeau's politics, as noted above, were similar to U.S. machine politics. In the 1950s, Drapeau was not interested in building expressways (other than the widening of Dorchester Boulevard). Expressways might disrupt neighborhoods populated by Drapeau supporters. Similarly, until the 1980s Drapeau was not an eager builder or supporter of public housing. Clark (1982) quotes Drapeau on this matter: "I have always favored housing development, but not necessarily *public* housing." In the U.S., machine cities were not overly eager to build public housing. Machine institutions (e.g., district elections) were highly conducive to political tradeoffs that kept public housing from being built (Bayor, 1988, 1989; Hirsch, 1983; Stone, Whelan, & Murin, 1986).

In summary, the Drapeau administration was, by and large, a caretaker regime. Drapeau practiced traditional machine politics and government comparable to those found in many U.S. cities in the same era (e.g., Chicago, New Orleans). Loyalty to the Drapeau organization was the highest virtue for citizens and councilors alike (Kaplan, 1982; Stone et al., 1986). Small property owners provided Drapeau's popular base, and they were given the kinds of policies they wanted (such as no public housing, favorable zoning decisions, no highways through their neighborhoods). Accommodations with major developers were made on a case-by-case basis in the *plans d'ensemble*.

At the same time, because of the "grand projects," Drapeau was a bit more than a mere caretaker. When Montreal acquired a "grand project," such as the World's Fair, then the city went on a crash program to build roads and subways. The result of this today is that the city has an excellent public transportation system. According to Colcord (1987), more than half the workers in the CBD come by public transportation; about one-third travel by car. In Boston, the figures are reversed. Mercer and Goldberg (1986) argue that the use and importance of public transportation is a crucial difference between U.S. and Canadian cities. While the subway and bus system are operated by the Montreal Urban Community we should note that the MUC receives substantial subsidies from the provincial government for the operation and construction of mass transportation.

Obviously, the "grand projects" approach had its advantages for Montreal. Events such as the World's Fair and the Olympics were important for Montreal's image as a world class city of the first rank. These events directly stimulated tourism, in both the short and long run. The building of urban infrastructure—subways, roads, and so forth—was a major benefit. There were other direct benefits from these "grand projects." Habitat, built by

Montreal architect Moshe Safdie as part of the 1967 Expo, is still a successful middle-income housing development today. Other Expo facilities are used for cultural and recreational facilities, although it is reasonable to say that the World's Fair site islands are underutilized today, and that the Expo was not the stimulus to downtown development that World's Fairs have been in other cities (Artibise, 1988; Dodd, 1988). The Olympic Stadium, now covered with a dome, provides a home for major league baseball's Expos. Other Olympic facilities, such as the pools and the mast above the dome, are tourist attractions.

At the same time, the "grand projects" approach had serious disadvantages. The overall lack of both planning and a strategy hurt the city badly when the economy declined in the 1970s. Manufacturing, transportation (including the port), and finance were the traditional bases of the Montreal economy. As early as the 1940s and 1950s, many manufacturers moved from the center of Montreal into outlying suburbs. This movement was accelerated in the 1970s by a depressed economy, and the movement of Anglophone firms to other provinces after the *Parti Quebecois* was elected at the provincial level in the mid-1970s. Moreover, the "grand projects" approach ignored the concerns of ordinary citizens, especially the poor. There was no large program of housing construction, and there was little attention to employment, retraining, and other programs for the unemployed. As a shrewd politician, Drapeau recognized these failures and redirected his administration in the 1980s.

THE SEARCH FOR SOLUTIONS

The Drapeau administration, for most of its years in office, may be viewed as parallel to the urban machines that dominated U.S. urban politics for many years. In most U.S. cities, these machines have been supplanted by new political forces: grass-roots community organizations, minority groups, municipal employees, and city bureaucrats. While the analogy is not perfect, there are distinct parallels with the rise of the Doré administration.

Doré's party, the Montreal Citizens Movement (MCM), was formed in 1974 as a coalition of grass-roots associations and individual citizens who were opposed to the Drapeau administration. These groups included Francophone unionists and professionals (the Parti Quebecois constituency), the Anglophone left, the trade-union coordinating body, and the local New Democratic Party. The MCM stood for decentralization and included neighborhood councils in its party platform. The MCM was not the only group that presented a challenge to the Drapeau mayoralty. Preservationists in Vieux Montreal and in groups such as Save Montreal fought to keep the city's rich architectural heritage (Gabeline et al., 1975).

The growth ideology, which seemed to be the motor for public and private strategies of North American and European cities, experienced few persistent criticisms (Molotch, 1976). In the course of time, the growth ideology was imposed naturally. Before the beginning of the 1970s, the outlines of an investigation, by city authorities or by business associations, on the nature and the urgency of intervention are very difficult to discover. The postwar years and the fulfillment of investments—each one bolder and more imposing than the other—seemed to have convinced the Montreal public authorities and the economic leaders of the inevitability of growth. It was enough to let their power be deployed in the Montreal area. All the more so because the charismatic power of Mayor Drapeau had proven its efficiency in attracting to Montreal, when necessary, the good graces of superior governments.

One study (Emard, 1985) of the postwar years shows well that thought on the present and future economy of Montreal was reduced to that very simple expression. It was admitted, if not ardently wished, that the city should lose its status as an industrial city in order to join the club of tertiary cities and be classified among world cities. That thought was equally characterized by a voluntaristic optimism which stipulated that all will go well when you wish it profoundly, and if you say it passionately. Also, any criticism that doubted the economic health of Montreal or pointed out the serious constraints to its future growth was interpreted as subversive behavior.

In spite of that, one such "subversive" attitude polluted the Montreal atmosphere at the end of the 1960s. A series of academic studies recognized, in effect, the weakness of the foundations of the Montreal economy in the midst of the Canadian and international economy. Moreover, these analyses raised new questions, more and more doubting of Montreal's capacity to continue playing a central role in the growth of Quebec and the Montreal region (Léveillée, 1978).

Some sort of reaction to the critics was to be expected. That reaction consisted of very positive expressions of opinion on the courage and the good historic fortune of Montreal against adversities, perceived or real. Colloquiums and public and private statements were the first steps in a strategy that culminated in a publicity blitz in the newspapers; in turn, the business leaders of Montreal came to project the smiling message of their determination to remain in Montreal and make their investments prosper. Conforming to their slogan, "La fierte a une ville" (Pride in the city), they tried to demonstrate their pride in Montreal and their wish to be associated with the movement.

It will always be difficult to measure the effects of these media campaigns against the events that followed, all the more so because the taking of power by the *Parti Quebecois,* the party advocating the separation of the province of Quebec from the rest of Canada, raised apprehensions, anxieties, and manipulation in certain Montreal businesses. Moreover, the end of the 1976 Olympic Games sounded a stop, for a period that all thought might be long, to the growth of public and private investments.

By the end of the 1970s, the Montreal public authorities had gone from a waiting attitude to interventionist attitudes and behaviors. The MCM, because of its ideology and its previous declaration, had to agree with this. From their side, businessmen accepted with good grace that the city would take over direction of a certain number of initiatives considering economic revival in Montreal. The business community had been asking for such a posture for many years (Villeneuve, 1984).

However, the business leaders were not ready for local political leaders to dominate the process of urban economic development. The business leaders knew that Montreal political leaders (including Drapeau) were not as influential as they were in the 1960s and before the 1976 Olympics. So, the business community decided to take a more active part in boosting Montreal and revitalizing Montreal. Instead of working behind the scenes, they reorganized their associations with the intention of being part of the public debate on the economic future of Montreal. As we will see below, Montreal business organizations have been involved in major public-private efforts in the past decade. Moreover, the Francophone business community grew tremendously in influence. The Chamber of Commerce is mostly Francophone and represents small and medium-size businesses. The Montreal Board of Trade historically represented the Anglophone big business community. Presently, the Board of Trade is divided between Anglophone and Francophone firms, with the rise of Francophone multinational companies such as Provigo, La Laurentienne, Bombardier, Lavalin, and SNC. The Anglophones let the French businessmen take the lead role because it was easier for them to get media coverage and because they were more aggressive. Moreover, the French entrepreneurs wanted to preserve their recent economic gains.

Since 1978 civic, political, and administrative leaders have tried to let their actions be guided by this spirit of public-private partnership. Cooperation and coordination are the key words in this relationship. For all actions that carry some consequences for the economic future of Montreal, private and public leaders are willing to build common strategies. Some battles are led by private personalities, while others are undertaken by public figures. Sometimes formal organizations that have been created since 1978 are used to implement strategies, while ad hoc structures are sometimes employed.

Officially, conflicts, contradictions, or even divergences don't seem to exist. The Civic Party, which governed until 1986, can be described as a conservative party. The MCM, which has been in power since 1986, can be described as a socio-democrat party. Under both, a consensual attitude and partnership with the business community are the rule.

After a long period in which the departure of industries had been perceived as a beneficial liberation of central space, the deconcentration of industrial jobs was now analyzed as a trend to reverse. In the same manner, the decentralization of commercialized investments into a network of suburban shopping malls was seen as a development that threatened to smother Montreal's neighborhood commercial streets over the long-term. Finally, the pro-suburban bias of the home building industry was seen as something that had to be changed in order to encourage middle-class families to remain in the city.

The search for solutions has been real since 1978. Private economic leaders were good analysts of Montreal's position in a regional, national, and international setting. In their actions, they emphasized the positive assets of the Montreal region instead of the economic constraints. In the process, they met a new brand of political and administrative leaders who were in power at City Hall. This was true of the Civic Party between 1978 and 1986, and it is true of the Montreal Citizen's Movement since 1986. During the past year, the MCM has been successful in its media campaign to show that the Montreal economy is in very good shape, even if the unemployment rate is high. Will business leaders continue to follow their boosterism campaign? Will neighborhood and community groups be convinced that the economy is in good shape? Will it be necessary to change the nature and the rhythm of the growth strategy that has been followed in cooperation with business leaders? Will it be possible to conceive a strategy that mixes tertiary sector development, emphasizing the central business district, with a strategy for the "other" society?

STRATEGIES TO REVITALIZE MONTREAL'S LOCAL ECONOMY

Since the end of the 1970s, Montreal's policies on revitalization can be subdivided into four components: manufacturing industries, neighborhood commerce, housing construction, and tertiary functions in the central business district. These policies were inspired by the objective of rebuilding the strongly threatened quality of life in the city.

Since 1980 the Montreal administration has been involved in the planning of nine industrial parks and the preparation of several sites for the development of 20,000 housing units. At the same time, the city began a program of revitalizing commercial arteries with the intention of making the "principal neighborhood streets" more attractive and more dynamic. According to the city's logic, people would remain in the city, or would return to it, thanks to the first two programs. The attraction of new residents would increase business in neighborhood commercial districts by 25% and would compete with suburban commercial malls.

However, the setting aside and planning of new sites was not sufficient for a successful policy of economic revitalization. Montreal administrators had to develop a more interventionist strategy to give public support to private initiatives. A series of direct financial aid programs was implemented: programs of loan guarantees and investments (PROCIM I and II); a program of financial subsidies for the acquisition of new houses, within the Operation 20,000 Houses program; and a program of financial and technical support for neighborhood merchant organizations (SIDAC).

The implementation of these initiatives represented a small administrative revolution at City Hall. Certainly the city of Montreal has always devoted a large part of its operating and capital budget to urban infrastructure. Between 1960 and 1978, the city's budget priorities emphasized major expenditures on infrastructure related to the subway, the 1967 Expo, and the 1976 Olympic Games. However, the preparation of sites for private ends, in particular the industrial sector and above all, direct financial assistance to particular industries to encourage them to invest in the city, were not regular practices at Montreal City Hall in the 1960s and 1970s. Municipal intervention of that sort had been "normal" practice at the beginning of the twentieth century. This had been abandoned over the years in favor of an unrestricted free market economy. The return to an interventionist philosophy was a revolution at City Hall for an administrative apparatus that was accustomed to a supervisory role instead of an entrepreneurial role.

It is not our purpose here to compare the last two administrations of the Civic Party (1978-1982, 1982, 1986) with the achievements of the Montreal Citizen's Movement since 1986. But suffice it to say that the MCM administration has made great efforts to implement an entrepreneurial spirit in City Hall administrative practices. This is noted in regard to the "localist" programs that characterized the reorientation of Montreal's administrations between 1976 and 1986. This is even more obvious in regard to new programs directed both toward tertiary functions in the central business district and toward a greater international role for the city (as we will see below). These last priorities were not undertaken in a systematic manner by political and administrative leaders before 1986, i.e, the Dore administration.

FROM INDUSTRIAL TO TECHNOLOGICAL PARKS

Since the end of World War II, suburban industrial parks have been available to industries that wanted to leave Montreal. And, for reasons we alluded to earlier, Montreal did not try to retain industries within its boundaries. At a point in the 1970s, faced with demographic decline and economic stagnation in all sectors, Montreal decided that something had to be done for the traditional work force instead of looking exclusively for new tertiary sector enterprises. All resources had to be protected. Unplanned affluence would not reach Montreal any more. Moreover, a number of analyses indicated that small and medium-size businesses generated more jobs than larger businesses in the Montreal metropolitan area, as well as in other metropolitan settings.

Since 1980, 10 successful industrial parks have been created. Certainly, that did not stop the process of deindustrialization, but it helped to minimize the consequences of the continuing departures of large industrial investors. The industrial parks strategy is still being implemented. The financial support programs were continued from the Drapeau to the Dore administration. The trend in favor of more sophisticated industrial enterprises has been accelerated. In the past few years, major efforts have been concentrated on the creation of technological sites that could attract research and development enterprises needing central locations and central communication networks. Traditional industries looked for traditional workers, skilled and unskilled. Tertiary industries need more skilled people with high levels of educational attainment. These are precisely the people who value residence in the central city and form the major part of the affluent society within Montreal.

It is rather ironic to observe that a "conservative" (Drapeau) administration initiated programs in favor of traditional industries and workers while a "socio-democratic" (Doré) administration insisted on more sophisticated enterprises and upper-class workers. However, we must add that the MCM administration is desperately searching for a strategy that could give new life to local or neighborhood economic development. Some analysts have commented negatively on the disparities of efforts on behalf of a neighborhood-based strategy when compared to efforts on behalf of a high-tech, CBD-based international city strategy. Community groups do not seem to attract as much attention as major business leaders when funds are distributed and new initiatives are undertaken. Ideas and projects coming from City Hall and from major business groups are said to be of a more general "public" interest than projects formulated for neighborhood ends. Moreover, the projects of the public-private partnership appear to be of more important consequence because they are big and "modern." Neighborhood-based efforts appear small-scale and traditional in contrast. The mass media publicize the ideas, projects, and campaigns of City Hall and the major business organizations while ignoring or forgetting community-based ideas and projects. Politicians can't ignore this situation, which is a fact of political life in most large cities.

FROM NEIGHBORHOOD TO CENTRAL COMMERCIAL DISTRICTS

Under the 1978 localist strategy, major investments were made to revitalize neighborhood commercial arteries. Streets were repaved. Sidewalk renovations that were undertaken provided more amenities. Neighborhood merchant associations were encouraged to exercise leadership and enhance local consumption.

This strategy is still being implemented, but the emphasis is not as great as it was in the later years of the Drapeau administration. No new neighborhood programs have been added, although the old ones have been continued. This appears ironic once again, because the MCM built itself as a grass-roots, neighborhood-oriented party. In recent years, the MCM acted as if it had to prove that the party was not exclusively oriented toward the neighborhoods and against the central commercial area. For example, in 1984 (two years before the MCM won control of City Hall), the party associated itself with business interests fighting for commercial development that suited local merchants, as opposed to potential extra-local investment. The 1984 controversy occurred over the proposal of an integrated commercial-cultural development on McGill College Avenue, in the central business district, between Ste. Catherine, Montreal's main shopping street, and President Kennedy. Cadillac Fairview, a large Canadian development corporation, made a deal with a reluctant Mayor Drapeau to incorporate a new building for the Montreal Symphonic Orchestra (with provincial government financing) with a big Eaton Center (Eaton's is a large Canadian department store chain). The scope of the project was such that McGill College Avenue would be blocked by "glass galleries."

Business investors on McGill College said the proposal was unfair and unacceptable. They had previously built their offices in conformance with an unofficial urban plan for the avenue. They respected the recommended width of the street to preserve the view of McGill University and Mount Royal. The Cadillac-Fairview project was a direct negation of the McGill College planning concept. Local merchants argued that such a huge commercial undertaking would kill their businesses on the street. The MCM decided to join with the local business community and other critical groups to fight

Drapeau and Cadillac-Fairview. This strategic alliance between the MCM and local business interests remained alive after the party came to power in 1986. Major commercial and office investments were made in the central district, without any opposition from the Doré administration.

A master plan for Montreal's central business and commercial district is being developed. The planning process has included two years of consultations, public hearings, and a final synthesis of discussions. In the meantime, critics are concerned about a number of disturbing trends. For example, the approval of the Marathon Realty project and the Prodevco-Lavalin 45-story office tower in the fall of 1988 seem to go against both the substance and the process of the proposed master plan (Wolfe, 1988). The consensus developed during the planning process has been stretched to the limit, or even contradicted, by decisions of this nature. The Central Business District plan is in its final phase of consultation and revision. In about a month, meetings took place in each of the eight *arrondissements* of the city, except the Central district, to make known the Synthesis for planning and developing these neighborhoods. The process was previewed, and the local populations expressed their demands and their opposition. The results of that exercise will furnish the chapters of the legislative agenda for the MCM administration in the course of the next four years. It is a cheap but also very interesting way of building an electoral and a political program.

It is, seemingly, not necessary to intervene in the Central district. Jobs and commerce are going relatively well; tertiary jobs are being created; commercial and service investments are made. In contrast, even with political and administrative programs, it is very difficult to stimulate local commerce and local jobs in many of Montreal's neighborhoods. Under these circumstances, what is the rationale for central district intervention? Some political and administrative leaders are forecasting that the current investment boom in the CBD will not last, or will become "artificial," if most Montreal neighborhoods are unable to get out of their current state of unemployment and relative poverty.

HOUSING A DUAL SOCIETY

Montreal city governments have not been active in the field of housing, except for the preparation of sites for private investments. In some older parts of the city, City Hall was happy to support the destruction of houses to make room for expressways, office buildings, or other investments. At the other end of the spectrum, they did not try to develop programs of public housing. Government's intention was to build a city for the white-collar workers of the tertiary sector. The departure of unskilled industrial workers was seen as an inevitable, if not desirable, result of the transition to a tertiary sector city.

That changed substantially when it became obvious that, despite its efforts, the city was losing its middle-class population while retaining its captive and poor populations. The Drapeau-Lamarre administration decided to create new programs to make the city more attractive for middle-class citizens. The original 1978 program was called Operation 10,000 Houses. It expanded its goals and became Operation 20,000 Houses in 1982, then was continued by the Doré administration after 1986. Under this program, the city sells or rents sites that it owns or that it has bought. The prices are competitive, and the first new buyer is subsidized for the amount of the down payment and the first five years' property tax contributions. It doesn't cost the city a lot, and all available sites are now occupied. In these respects, the program has been successful.

In the past two years, the MCM administration has had trouble finding a strategy that could parallel Operation 20,000 Houses and also be directed toward the less-affluent society. Governmental subsidies in the field of social housing have been cut, and the city administration is unable to respond to the growing needs alone. Public hearings have been held in order to find a compromise that could become an acceptable housing policy. A final report has been published, but no action has followed. Public money is so restricted that it seems almost impossible to do two things at the same time: that is, to implement programs that could attract middle-class people, and to finance new programs of social housing. So little has been done in public housing in the past that substantial action in the field of social housing is like acting anew. Moreover, due to the increasing rate of unemployment and poverty over the past 10 years, the needs for social housing are higher than ever.

THE IMPACT OF REVITALIZATION

Montreal's role in the Canadian, continental, and global economies has diminished considerably since the 1950s. There are many who are not satisfied with a lesser role for Montreal. Higgins (1986), observes that Montreal "operates in an economic space that is ultimately world-wide." The Montreal Chamber of Commerce has encouraged the development of Montreal as an international city (Léveillée, 1988). In November 1988 a symposium was held in Montreal on the topic "Montreal, international city?" Montreal Mayor Jean Doré spoke of his vision of Montreal as the first city of "American-Europeanism," the principal port of entry for Europe in North America in the years following both the U.S.-Canadian free trade pact of 1988 and the 1992 end of trade barriers in the European economic community (Pare, 1988). In a recent publication aimed at the investment community,

Montreal compares itself favorably to 13 other major international cities: Toronto, Vancouver, New York, Boston, Chicago, Atlanta, Los Angeles, Paris, Milan, London, Frankfurt, Stockholm, and Tokyo (Communuate Urbaine de Montreal, 1989). Montreal also belongs to an international organization of larger cities that discusses common problems, such as transportation.

International activities are clearly an economic development priority for the Doré administration. A 1986 federal commission (the Picard commission) recommended seven areas of concentration in the international economy: headquarters of international corporations, high technology, finance and international trade, design, cultural activities, tourism, and transportation. Another consultant's report to the Doré administration called for the city to become a true international city, with headquarters of international organizations, international conferences, and the establishment of an international trade center. Capping all these will be the establishment of the Canadian Space Agency in the Montreal area (Levine, 1989). Still, it is more likely that Montreal's future role will be as a "regional nodal diversified service center," with its base in Quebec province and parts of the northeast United States.

Two extremely well-informed analysts, Hero and Balthazar (1988), argue that as Montreal lost its Canadian role, it gained in relative importance to Quebec province. They note greater economic links with the United States after the opening of the St. Lawrence Seaway, and greater cultural, communications, and commercial linkages with Europe. Hero and Balthazar point out that "by the early 1980s, Greater Montreal and its nearby dependent environs accounted for nearly three-fifths of Quebec's population, almost two-thirds of its manufacturing jobs, and four-fifths of its financial, business and professional services" (p. 403). Despite the decline in the port, they argue that Montreal's strategic location, on a major navigable river (the St. Lawrence) within 500 miles of the continent's population and industrial base, inevitably makes it one of the top North American ports. Depending on the method of counting, Montreal still ranks with such ports as New York and Baltimore. Hero and Balthazar further argue that the city's cultural diversity and cosmopolitan nature give Montreal great potential as a "bridge between Anglophone North America, the French-speaking world, and . . . Western Europe" (p. 403).

The city's changing economic role is reflected in different employment patterns. The tertiary sector grew as the industrial sector of the economy declined. In the 1981-1986 period, the growth in tertiary employment was remarkably similar throughout the region. But, it is clear that the central business district plays the pivotal role in such areas as hotel, office, and financial employment (Lamonde et al., 1988). The increases in the tourism-related category provide an interesting comparison with U.S. cities. Some

U.S. cities have been notably more successful than others in attracting tourist and convention business. Undoubtedly, the successful tourism and cultural programs of the Quebec provincial government aided the growth of tourism. The growth in the high-technology-related portions of the third sector is helped by the presence of several excellent universities in the Montreal area, including McGill, University of Montreal, and University of Quebec at Montreal (Artibise, 1988).

At present, there is a great deal of vigor in downtown construction. Several new projects opened in 1988, and others are in progress. These downtown projects include the Centre Eaton, which has 210 shops and 310,000 sq. ft. of retail space; the Place Montreal Trust, which includes 420,000 sq. ft. of retail space and 580,000 sq. ft. of office space; the Maison des Cooperants, which includes 160,000 sq. ft. of retail use and 500,000 sq. ft. of office space; and Les Cours Mont-Royal, which includes 200,000 sq. ft. of retail space with 150 shops, 300,000 sq. ft. of office space, and 150 condominiums (Smith, 1987).

Changes in the city's employment structure and the continued expansion of the central business district have obvious implications for the city's social and economic geography. As Harold Kaplan (1982) observed: "Montreal is not one community but at least two. St. Lawrence Boulevard or Boulevard Saint-Laurent has been the traditional line dividing the Francophone east from the English west." Until very recently, the Anglophone group was strengthened by the addition of immigrants from Ireland, Italy, and eastern Europe. Kaplan (1982, p. 312) notes that "the French and English were not merely different language groups, occupying different parts of the city. They were entirely separate social communities, practicing different religions, attending different schools, living in different cultural worlds, and forming stereotyped images of the other side."

The Anglophones, for many years, possessed the dominant portions of economic power. This was reflected in the development of the central business district to the west. As Limonchik (1982) points out, one of Mayor Drapeau's objectives was "to achieve a more balanced distribution of economic and political power between the English- and French-speaking parts of Montreal by directing investment to the eastern part of downtown." This shift was attempted in a number of ways: the location of the main Metro station (Berri-Demontigny) in the east end of the city, and the location of numerous public facilities, such as the convention center, the arts center, the Complexe Desjardins, the Office of Hydro-Quebec, and the University of Quebec at Montreal campus (Barcelo, 1988).

In 1981, slightly more than one million people lived in the city of Montreal—62% of these were of French origin, 9% of British origin, and the rest of other ethnic backgrounds (Marois, 1988). There are notable socieconomic disparities in the city. The west side is predominantly English, and the east side is overwhelming French. Generally speaking, the west side is more affluent than the east side, although some west side neighborhoods, such as Pointe-St-Charles, are similar to poor east side neighborhoods, such as Hochelaga and Maisonneuve. There have been significant changes in immigration patterns. First of all, there has been a substantial reduction in the numbers of immigrants entering Canada. Second, in Montreal's case, European countries (Italy, Greece) were the main source of immigrants for many years. From 1976 to 1981, the two largest sources of immigrants were Haiti and Indochina. (Quebec immigration rules give preference to French-speaking immigrants.) After the adoption of the French language charter in 1977, new immigrants were directed into the French school system (Marois, 1988).

Marc Levine's work is highly suggestive on how these social and linguistic changes affected the Montreal economy. Anglophones dominated the city's economy through the 1960s. Provincial policy attempted to promote the economic interests of Montreal Francophones in four ways: direct and indirect employment generated by government activity, promotion of a Francophone managerial and business class, public investment in education, and direct public regulation of the language practices of private corporations (Levine, 1988). Francophone firms such as Bombardier, Provigo, and Lavalin achieved Canadian and international scope in the 1970s, although Anglophone firms, such as Molson, the Bank of Montreal, and Canadian Pacific are still prominent in the Montreal economy. The economic position of Francophones in Montreal has improved substantially in the past 25 years in terms of wages and gains in managerial position, and Levine concludes that "through political mobilization and public policy and with help from some favorable market trends—the historically disadvantaged Francophone majority made remarkable progress in a short, 25-year period. Montreal's economy may not function completely *en francais* and Francophones may not be *maitres chez nous*. But, in the city's economic affairs, the days of the "'English city' are clearly over" (pp. 60-61).

REFERENCES

Artibise, A. F. J. (1988, Fall). Canada as an urban nation. *Daedalus, 117* (4), 237-264.

Barcelo, M. (1988). Urban development policies in Montreal, 1960-1978: An authoritarian quiet revolution. *Quebec Studies 6*, 26-40.

Barnekov, T., Boyle, R., & Rich, D. (1989). *Privatism and urban policy in Britain and the United States.* New York: Oxford University Press.

Bayor, R. H. (1988, November). Roads to racial segregation: Atlanta in the twentieth century. *Journal of Urban History, 15* (1), 3-21.

Bayor, R. H. (1989). Urban renewal, public housing and the racial shaping of Atlanta. *Journal of Policy History, 1* (4), 419-439.

Clark, G. (1982). *Montreal: The new cite*. Toronto: McClelland and Stewart.

Colcord, F. C., Jr. (1987). Saving the center city. In E. Feldman and M. Goldberg (Eds.). *Land rites and wrongs: The management, regulation and use of land in Canada and the United States*. Boston: Oeigeschlager.

Commission du Developpement Economique De Montreal. (1989, June). *Rapport sur l'economie locale* (Report on the local economy). Report submitted to municipal council.

Communaute Urbaine de Montreal. (1989). *Decision: Montreal*. Montreal: Communaute Urbaine de Montreal, Office de l'expansion economique.

Dodd, J. (1988). *World class politics: Knoxville's 1982 World's Fair redevelopment and the political process*. Salem, WI: Scheffield.

Emard, C. (1985). *La commission d'initiative et de developpement economique de Montreal* (Montreal's commission on social initiatives and development). Montreal: department de science politique, Universite du Quebec a Montreal.

Gabeline, D. et al. (1975). *Montreal at the crossroads*. Montreal: Harvest House.

Gagnon, A. G., & MONTCALM, M. B. (1989). *Quebec: Beyond the quiet revolution*. Scarborough, Ont.: Nelson Canada.

Gappert, G. (1988). *Montreal's urban form: Add planning to vision*. Paper presented at meeting of American Council on Quebec Studies, Quebec.

Hero, A. O., Jr., & Balthazar, L. (1988). *Contemporary Quebec and the United States, 1960-1985*. Lanham, MD: Center for International Affairs, Harvard University and University Press of America.

Higgins, B. (1986). The rise—and fall? of Montreal: A case study of urban growth, regional economic expansion and national development. Moncton, N. B.: Canadian Institute for Research on Regional Development.

Hirsch, A. R. (1983). *Making the second ghetto: Race and housing in Chicago, 1940-1960*. Cambridge: Cambridge University Press.

Kaplan, H. (1982). *Reform, planning and city politics: Montreal, Winnipeg, Toronto*. Toronto: University of Toronto Press.

Lamonde, P., et al. (1988). *La transformation de l'economie montrealaise, 1971-1986* (The evolution of Montreal's economic structure). Montreal: INRS—Urbanisation.

Léveillée, J. (1978). *Developpement urbain et politiques gouvernmentales urbaines dans l'agglomeration montrealaise, 1945-1975* (Urban development and urban politics of government in the Montreal metropolitan region). Montreal: Societe canadianse de science politique, Etudes en science politique.

Léveillée, J. (1988, Autumn). Pouvoir local et politiques publiques a Montreal: Renouveau dans les modalites d'exercice du pouvoir urbain. *Cahiers de Recherche Sociologique, 6* (2), 37-63.

Levine, M. V. (1986). Language policy, education and cultural survival: Bill 101 and the transformation of anglophone Montreal, 1977-1985. *Quebec Studies, 4*, 3-27.

Levine, M. V. (1988). The reconquest of Montreal: Public policy, language, and economic change, 1960-1987. *Quebec Studies, 6*, 41-64.

Levine, M. V. (1989). Urban redevelopment in a global economy: The cases of Montreal and Baltimore. In R. V. Knight & G. Gappert (Eds.), *Cities in a global society* (pp. 141-152). Newbury Park, CA: Sage.

Limonchik, A. (1982). The Montreal economy: The Drapeau years. In D. Roussopoulos (Ed.), *The city and radical social change* (pp. 179-206). Montreal: Black Rose.

Marois, C. (1988, Spring/Summer). Cultural transformations in Montreal since 1970. *Journal of Cultural Geography, 8* (2), 29-38.

Mercer, M. A., & Goldberg, S. (1986). The myth of the North American city: continentalism challenged. Vancouver, B.C.: University of British Columbia.

Molotch, H. (1976, September). The city as a growth machine. *American Journal of Sociology, 82* (2), 309-332.

Nader, G. (1976). *Cities of Canada: Vol. 2. Profiles of fifteen metropolitan centres.* Toronto: Mcmillan of Canada.

Noyelle, T. J., & Stanback, T. M., Jr. (1983). *The economic transformation of American cities.* Totowa, NJ: Rowman and Allanheld.

Pare, I. (1988, November 23). Jean Doré reve a l'amerocuropeene (Jean Doré dreams of American-European unity). *Le Devoir* (Montreal), p. 3.

Sancton, A. (1983). Montreal. In W. Magnusson & A. Sancton (Eds.). *City politics in Canada* (pp. 58-93). Toronto: University of Toronto.

Smith, K. (1987, June 13). They're racing to change how downtown looks. *The Gazette* (Montreal), p. H1.

Stone, C., Whelan, R., & Murin, R. (1986). *Urban policy and politics in a bureaucratic age* (2nd ed). Englewood Cliffs, NJ: Prentice-Hall.

Villeneuve, P. (1984). Rapport d'etape sur les groupes d'interet a Montreal au cours des annees 70 (Report on interest groups in Montreal during the 1970s). In J. Léveillée et al. *Materiaux por l'etude du systeme politique montrialais* (Materials for the study of the Montreal political system). Montreal: department de science politique, Universite du Quebec a Montreal.

Wolfe, J. (1988, October 22). Political will to save the Queen's Hotel was lacking. *The Gazette* (Montreal).

9

Regeneration and Quality of Life in Vancouver

WARREN MAGNUSSON

VANCOUVER'S ECONOMIC POSITION

Vancouver is one of the cities that hopes to profit from the decline of the older manufacturing centers. Unlike them, it is on the Pacific Rim, looking west to Japan and the Little Dragons of East Asia. Relatively unencumbered by the detritus of the first and second industrial revolutions, and well positioned between the new and the old economic powers, Vancouver appears to its own elites as a city of the future—Canada's western metropolis, Los Angeles to Toronto's New York. Hidden in this vision is a darker reality—a city buffeted by the destruction of its own industrial base, isolated from the centers of power in North America, and fearful of a future that could leave it as a provincial backwater.

Vancouver's position in the old global economy was at the other end of the productive system from the port cities and manufacturing centers of the Atlantic Basin. It had been founded in 1886 by Montreal entrepreneurs who had—with capital from London and the protection and support of the federal government of Canada—built the Canadian Pacific Railway to open up the British west, north of the American border. Vancouver was the Pacific terminus, the link to Asia, in the all-British route across North America from Liverpool to Hong Kong and Singapore. Once established, it quickly became the business center for the vast, mountainous province of British Columbia. Vancouver businessmen—some independent entrepreneurs, many agents of outside capital—organized the exploitation of B.C.'s forests and coastal waters, opened mines, and brought orchardmen and ranchers into the mountain valleys and plateaus of the south. Especially after the opening of the Panama Canal in 1914, the Port of Vancouver became the outlet for the farms, forests, and mines of Western Canada, which were producing food and raw materials for the great manufacturing cities of the Atlantic Basin (Evenden, 1978).

For the white men who came and overwhelmed the native population, British Columbia was, from the beginning, a rich province, a place of high profits and high wages, and Vancouver was at the center of its economic system. The years after the Great Depression were especially prosperous, as new technology enabled the lumbermen to strip the forests at an astonishing rate and feed the immense American demand for timber. As in Pittsburgh and Detroit, powerful unions emerged to take advantage of the prosperity of the basic industries—and extract wages appropriate for aristocrats of labor. The struggles over the division of surplus value were often intense, but the workers secured for themselves the highest average wages in Canada. No great sense of security came with this prosperity, though, because the province's resource industries were so obviously dependent on demand in the centers of the global economy.

Vancouver's position has been like Canada's in miniature: a place at the margins, a colony in the imperial system, a hinterland of the metropolis—yet, peculiarly favored in its cultural, political, and economic relations with the center. Political economists have disagreed sharply in their interpretations of Canada's position, some suggesting that it is really a weak, underdeveloping country, like Argentina or Uruguay, others placing it in the middle rank of "imperial" powers (Williams, 1988). On all accounts, Canada's "staples dependency"—first theorized by Harold Innis in the 1920s and 1930s—has rendered it peculiarly vulnerable to fluctuations in the global economy and given it little capacity to control its own development. Historic dependence on foreign capital for development, foreign domination of the resource and manufacturing industries, and the deep foreign penetration of economic and political elites are all symbolic of the country's subordination, first to Britain and then to the United States. From Vancouver's perspective, these relations are replicated in B.C.'s position as a colony or hinterland of central Canada. If Canada in general is a prosperous but vulnerable dependency, B.C. (and Vancouver) share in the prosperity, but are doubly dependent and doubly vulnerable.

THE CONTENDING PHILOSOPHIES

Vancouver was hit very hard by the recession and fall in commodity prices in 1981-1983 (Allen & Rosenbluth, 1986). Its experience serves as a reminder that the collapse of the old manufacturing industries in the Atlantic Basin has affected the resource producers in the hinterlands even more dramatically than the industrial producers in the metropoles. Of course, nothing in Vancouver can compare to the hardship in Latin America or Africa, but the effects of the recession in the center of a richly favored hinterland give a sign of what happened in less-favored areas. Official unemployment in metropolitan Vancouver more than doubled in 1981-1982, cresting upward to reach a peak of about 15% in the winter of 1984-1985, and staying well above 10% until 1987 (Vancouver, Economic Development Office, 1987). "Food Banks" sprang up across the city to relieve the suffering, and most of them remain today. Although the local economy has "recovered," real average wages are still below 1981 levels and official unemployment is more than 7.4% (Howlett, 1990). A general recession in the 1990s will probably double the latter figure.

Like New Zealand—another prosperous but staple-dependent region of the industrialized world—British Columbia has had a government determined to force an adjustment to the "new reality" of the global economy by cutting back on the public sector and exposing people to the storms of international competition. However, the B.C. government owes no allegiance to the labor movement, and it has been faced, like Thatcher in Britain, by strong opposition from the unions and other defenders of the public sector. The government's "restraint" program brought the province to the verge of a general strike in November 1983, and led to intense conflict with municipalities, school boards, and other autonomous public agencies over the next three years (Howlett & Brownsey, 1988; Magnusson et al., 1984). The symbolic culmination of the provincial-local conflict was the dismissal of the Vancouver school board (for refusal to adopt a budget within provincial guidelines) and imposition of a provincial trusteeship in May 1985. Since 1986 political tensions in the province have eased—partly because provincial tax revenues have improved, and the government feels less need to squeeze the authorities it funds—but the basic direction of the provincial government has not changed. British Columbia continues to stand out in the Canadian context as the place where the neoconservative policies associated elsewhere with Reagan and Thatcher have been pursued with greatest vigor.

At the same time, Vancouver is one of the few cities in Canada to have had a "progressive" civic administration. Municipal politics in Canada—unlike federal or provincial politics—is generally nonpartisan (Magnusson & Sancton, 1983). Only in Montreal, Quebec, and Vancouver has there been anything approaching a "party system" in recent years. In each case, the parties are purely local formations that align themselves roughly on a left-right spectrum. Since 1937 the dominant "party" in Vancouver has been the Non-Partisan Association (NPA), a good example of a "growth coalition" (Gutstein, 1983; Tennant, 1981). It was displaced in 1972-1976 by The Electors' Action Movement (TEAM), a liberal formation dedicated to growth control. And then, in 1982-1986—coinciding with both the recession and the province's restraint program—the council came under the control of an alliance of the left. Mayor Mike Harcourt, a New Democrat (now, indeed, provincial leader of the New Democratic Party), coalesced with aldermen from COPE, the Committee of Progressive Electors (an organization backed by the Vancouver Labour Council, with strong organizational input from the provincial Communist Party), to form a "progressive" majority on most issues—one strongly opposed to the provincial government and committed to quite a different vision of the policies necessary for adjustment to the "new reality" of the global economy.

In appearance, at least, there have been contending strategies for economic renewal in Vancouver: a neoconservative one advanced by the provincial government, and a social democratic one associated with the city of Vancouver. Both have been articulated in the context of a perceived crisis in the provincial and, hence, the municipal economy. Both have addressed themselves to the decline in the traditional resource industries and the opening of new economic opportunities for the city on the Pacific Rim. There is a common perception that Vancouver can somehow make use of its cultural connections with the Asian Pacific countries, and capitalize on its strategic location between Tokyo and New York, to become an important (if still secondary) control center in the global economy. The ambition, clearly, is to free the city from its dependence on the provincial resource base and ultimately to make it a relatively autonomous center of global capital—connected to Hong Kong, Singapore, Tokyo, Los Angeles, and Toronto, but subordinate to none of them. Given this common ambition, it is perhaps to be expected that neoconservative and social democratic strategies have ultimately converged.

THE POLITICAL CONTEXT

As steel was to Pittsburgh, so forestry was to British Columbia (Marchak, 1983). The crisis in the industry in the early 1980s was fully comparable to the contemporary crises in manufacturing. In the logging and mill towns of the interior, the effects were devastating—unemployment rates of 30% to 40% were common, and some communities became virtual ghost towns—but the reverberations were felt also in Vancouver, where the big forest companies are headquartered. The companies made it clear that their downsizing was permanent: They could neither maintain their present work forces nor keep wages and benefits high if they were to compete with offshore producers. The mining companies, faced with a dramatic decline in prices for their products, said similar things. The talk was of structural change, rather than cyclical downturn: an end to the province's high wage, high employment resource economy. For Vancouver, this was as frightening as for the province as a whole.

By global standards, Vancouver was and is still a comparatively small place. About 1.5 million people now live in the metropolitan area—less than one-third of them in the central city. Nonetheless, it is by far the biggest city in Western Canada, and almost of "major league" status by American standards. (Greater Vancouver is the sixth-largest metropolitan area on the Pacific Coast of the U. S. and Canada.) Its position in the province is absolutely unrivaled: About half the people in British Columbia live in the region. This is mixed blessing for the municipal authorities because it means that the provincial government is bound to take a close interest in the city. It is also a mixed blessing in the sense that the immediate hinterland for the metropolis is comparatively sparsely populated. This—combined with the Vancouver's isolation from other centers of population on the continent—has meant that manufacturing industry always has been comparatively underdeveloped.

The big companies headquartered in Vancouver are generally in the resource sector, especially forestry (Vancouver, Economic Development Office, 1989). The Vancouver Stock Exchange, founded in 1907, always has specialized in raising venture capital for resource exploration and exploitation. Historically, Vancouver has been a secondary financial center, dependent on Montreal and more lately on Toronto, where the big Canadian banks and financial corporations are headquartered. Since the 1960s, however, a small independent financial sector has emerged, with encouragement from the provincial government. The Hong Kong Bank recently took over the biggest of these independent financial institutions, the Bank of British Columbia. This occurred with the blessing of the provincial government, which was and is eager to both secure a base for Asian capital in Vancouver and offset Toronto's influence within the city.

Historically, the Board of Trade has been the most important of the formal business organizations in the city. The Downtown Business Association is also significant. More recently the "bilaterals," like the Hong Kong-Canada Business Association, have begun providing an organizational framework for the new immigrant entrepreneurs from Asia. The latter are a new, more internationally oriented element in what, historically, has been a very provincial business community. The Liberals and Conservatives have provided a framework for business participation in electoral politics federally and, until 1952, did the same provincially. However, in that year, the populist Social Credit Party came to power in an upset victory and soon positioned itself as the preferred party of "free enterprise" (Blake, 1985). Although relations between Social Credit and Howe Street (the proverbial business center) have been strained from time to time—the party appeals especially to people in the provincial interior, where no love is lost for Vancouver—it is clear that Vancouver business prefers Social Credit to the "NDP socialists."

To think of the NDP as socialist is to recall its origins in the old Cooperative Commonwealth Federation, which was established in 1932 as an amalgam of various farm, labor, and socialist groups. In British Columbia, the CCF had a particularly strong base in the labor movement, which had a long tradition of militancy. By 1941 it had emerged as a powerful electoral threat to the established parties—strong enough to impel them to coalesce against it. The CCF always was social democratic, not socialist, and its reconstitution in 1961 as the New Democratic Party (NDP) was intended to make that clear. The hegemony of the NDP on the left in provincial politics has long been unquestioned. Within the labor movement, the situation is much the same, although the Communists—who fought the CCF for control a generation ago—remain significant.

In some ways, Vancouver city politics mirrors provincial politics. The CCF began to contest aldermanic seats in the 1930s, and the Non-Partisan Association (NPA) was formed in response in 1937, to nominate and support non-socialist candidates for municipal office. The CCF never had the success locally that it did provincially, and the NPA dominated municipal politics until 1972. This was partly due to a system of at-large elections, which made it difficult for opposition candidates (and non-incumbents) to get elected to council. It was also an effect of the strong traditions of nonpartisanship in Canadian municipal politics. When the NDP attempted to re-enter municipal politics in 1970 after a long period of inactivity, it failed abysmally and had to leave the field to TEAM (The Electors' Action Movement) and COPE (the Committee of Progressive Electors), two purely local formations established in 1968. TEAM has faded, but COPE still forms the main opposition to the NPA.

It is important to keep in mind that, by comparison with American states, Canadian provinces are extraordinarily powerful. Their strength is partly the result of the cabinet-parliamentary system; provincial premiers and their cabinets are in command in a way that American governors cannot be. Party discipline is strict—which is not the case for the civic parties in Vancouver—and the governing party normally has a secure majority. Beginning in the 1950s the provinces underwent a period of administrative modernization that enabled them to intervene more systematically in matters of social and economic policy. Increasingly, they claimed prime responsibility for the whole range of social services (health, education, welfare, recreation, culture) as well as for the development of the provincial economies. Matters once left to municipal or other local authorities became subject to provincial action, and the provinces jealously guarded their jurisdiction against the federal government.

In B.C., as in other provinces, there has been some pressure to consolidate local authority, but the local state in Vancouver remains highly fragmented. Between 1914 and 1936 metropolitan special-purpose authorities were created for water, sewerage, and public health. They were brought together in 1965, when the province established a system of regional districts for inter-municipal administration (Tennant & Zirnhelt, 1973). The governing board of the Greater Vancouver Regional District (GVRD) is composed of representatives of the constituent municipalities (and unincorporated areas). The GVRD has certain mandated functions, but otherwise performs tasks delegated by the municipalities from a list specified by the province. Until 1983 it did regional planning, but in that year the province stripped all the regional districts of the their planning functions, ostensibly to streamline the approval process for developers and to protect municipal jurisdiction. Throughout the province, school boards are elected independently of the municipal council, and in the city there is an independently elected Parks Board. Public utilities, public transport, social services, income assistance, and health services are all provided by either provincial agencies or agencies under direct provincial regulation. Thus, the municipal council is even more exclusively focused on issues of urban development and planning than its American counterparts.

The federal government has been retreating from urban affairs since the 1970s. It had become a major actor in 1944, upon establishment of the Central (now Canada) Mortgage and Housing Corporation (CMHC), the Canadian equivalent of the FHA. The CMHC underwrote the postwar suburban expansion and was involved (in imitation of its American counterpart) in urban renewal and public housing. Ottawa took some of the blame when complaints developed in the 1960s about the effects of downtown redevelopment, suburban sprawl, and ghettoized public housing. It reformulated its housing programs in 1973, to encourage more mixed land-use, and began to use abandoned harbor-front or railway lands to create showpiece developments in many of the major cities. However, a long period of fiscal restraint began in Ottawa in 1975, and the federal government generally has not been prepared to make major commitments to housing and urban development. This has been rationalized in terms of a renewed respect for provincial jurisdiction in these matters. Thus Vancouver has been left almost entirely to the mercies of the provincial government since the late 1970s.

COMPETING STRATEGIES FOR THE CITY

Until recently the province's development strategy for Vancouver could have been best characterized as laissez-faire. During the first two decades of Social Credit government (1952-1972), the focus was very much on the B.C. interior, where the Party had its strongest base of support. Although it regarded itself as a "free enterprise" party, Social Credit did not shy away from spending on roads, dams, and other facilities it thought necessary for the development of B.C.'s natural resources. It even brought the hydroelectric industry under public ownership to facilitate its expansion. The showpieces of the government's strategy for economic development were its resource megaprojects, like the Columbia River hydroelectric development. In this context, Vancouver and its suburbs were subject to a benign (or not-so-benign) neglect, which allowed private developers and municipal councils to proceed much as they pleased, relying as necessary on support from Ottawa's Central Mortgage and Housing Corporation.

After the Second World War, the Vancouver council followed policies much like the ones adopted by American cities in the same period. Indeed, its Master Plan and Zoning By-Law had been developed in 1926-1929 by the American planner Harland Bartholomew and revised by him (to include a proposed freeway system) in 1947. The small council and at-large voting system, established in 1936, followed a familiar American model, and the city's administrative system was reorganized following a report from Chicago's Public Administration Service in 1955. The post-war council focused on

> the promotion of physical growth and development. This was to be achieved through a series of interrelated activities: downtown revitalization by public and private initiatives, higher-density development in residential areas, publicly funded urban renewal to promote industry, and a system of freeways to tie all parts of the city together [Gutstein, 1983, p. 199].

The recession of the late 1950s combined with increasing anxiety about the drift of population (and industrial and commercial activity) to the suburbs to increase the pressure from organizations like the Downtown Business Association for vigorous growth promotion. Again, in 1963, council turned to American consultants "who proposed building a system of freeways, encouraging apartment construction near the city center (to provide shoppers and a pool of labour), and assembling land for commercial development with public funds and turning the land over to private interests for redevelopment"

(Gutstein, 1983, p. 200). Efforts in this direction coincided with the long boom of the 1960s.

Eventually, these policies produced a reaction. The urban freeway proposals proved unpopular (Leo, 1977), and the rapid high-rise apartment development in the West End (between the Central Business District and Stanley Park, the city's major park) heightened anxieties about the deteriorating quality of life in the city. In Vancouver, as in other parts of Canada, the perceived unlivability of American cities—dramatized by the contemporary ghetto rebellions, but remarked already by Canadian tourists—was a point of reference for political debate. The fact that the council was following American models led critics to suggest that by adopting measures that had failed in the United States, it was devastating a city that still worked. It was in this context that TEAM and COPE formed, in 1968, in opposition to the NPA. In 1972, TEAM, the milder, more middle-class reform group, won an overwhelming victory in the municipal elections and took control of the mayoralty, the council, the school board, and the parks board.

The TEAM-dominated council attempted to open up planning to greater citizen participation, moderate the pace of development, reduce heights and densities, discourage downtown traffic, and encourage projects that would contribute to urban vitality. As in Toronto, Jane Jacobs' ideas about mixed urban uses proved influential. The False Creek and Granville Island developments, south of downtown, were of great symbolic importance for the medium densities, and the physically, socially, and economically mixed character of these areas was in stark contrast to the "high-rise desert" of the West End (Ley, 1980, 1981). Both these developments were supported by the federal Liberal government, which was desperately attempting to restore its political base in Western Canada. Mayor Art Phillips of TEAM emerged as the great Liberal hope in Vancouver, and the Council under his leadership could count on considerable federal support. This was a propitious time for the reformers provincially as well, for the New Democrats won their first (and only) provincial election in 1972. This meant that, for the only time in the province's history, it had a government without close ties to the Vancouver business community, and hence to the interests that supported rapid downtown development. The NDP government put tough controls on the conversion of agricultural land to other uses. The effect of this was to inhibit suburban sprawl. At the same time—and partly in response to this policy—the Greater Vancouver Regional District developed a strategy for growth that focused on the creation of suburban "town centers," which would relieve the pressure on downtown. For the city of Vancouver, which enjoyed substantial representation on the Board of the GVRD, this was clearly a complementary policy. The new sensibility embodied in these policies was symbolized by the title of the GVRD's 1975 Plan: "The Livable Region."

These policies of growth restraint reflected a shift in the social base of Vancouver, which in turn reflected its development as a corporate headquarters center. With a larger urban middle class came demands for greater urban amenities and more meaningful citizen participation in growth planning. However, the global stagflation that became apparent by the mid-1970s ultimately increased anxieties about the city's economic future. Social Credit returned to power provincially at the end of that year, and—although it made few dramatic moves during its first term—it talked of deregulating business and reducing public spending. In Vancouver, Art Phillips retired from the mayoralty, TEAM split, and a coterie of traditional, business-oriented politicians regained control of City Council in 1976-1980. This set the stage for a series of big downtown development projects.

The opportunity for these projects was provided by the relocation of port facilities, railway yards, and related industrial activities near the CBD. Some of the land belonged to the National Harbours Board (and hence to the federal government), but the key properties were owned by the railways, especially the CPR. After some unsuccessful efforts to develop the lands itself, the CPR's real estate arm sold its vast holdings on the north shore of False Creek to the provincial government. This gave the province direct control over the key development site in the city, and the opportunity to create its own urban showpiece, to rival the one developed by the TEAM-dominated City Council and its federal Liberal allies (Ley, 1987). The city and the province might have worked together on "B.C. Place," had the conservatives remained in control at the municipal level. However, provincial-municipal relations grew frosty when New Democrat Mike Harcourt captured the mayoralty in 1980, and frostier still when the political balance on the council shifted decisively to the left in 1982, just at a time when the provincial government was moving firmly in the other direction. The province was not about to share credit with its political enemies.

The first of the great projects at B.C. Place was a domed stadium, of the sort that is supposed to attract baseball franchises. The second was Expo '86, originally intended as a transportation fair to celebrate the one-hundredth anniversary of the city. Expo offered the excuse for rapid development of an Automated Light Rapid Transit (ALRT) system to bring visitors to the fair. It also legitimated the use of public funds to clear a huge tract of land near the CBD for private redevelopment (Anderson & Wachtel, 1986). The two projects were nicely timed to precede the provincial elections of 1983 and 1986, in both of which Social Credit was successful. Expo proved particularly popular with the public, and afterwards its site was sold to Hong Kong billionaire Li Ka-Shing at a price not much greater than the cost of cleansing the soil (Gutstein, 1988; Matas, 1989).

As various critics pointed out, the province was applying the megaproject development strategy, which it had used for decades in the interior, to the City itself. Expo was the key to the strategy: the focus for an effort to attract tourists and new investors. Both the tourists and the investors were supposed to come from the Pacific Rim, especially California, Japan, and Hong Kong. This idea reflected the growing consensus in the business community that Vancouver's future depended on its development as a center for Pacific Rim trade and commerce. The new orientation to the Pacific Rim provided a further rationale for the government's neoconservative policies. It was claimed that Asian and American investors, accustomed to operating in nonunion environments, would balk at coming to B.C., with its powerful and militant unions. These potential investors also would object to the large public sector in B.C. and the high levels of taxation that went with it. Thus, the provincial government was able to frame its 1983 restraint initiatives—measures to reduce spending and employment in the public sector, weaken the unions, reduce social rights, and generally free private business from external controls—as competitive necessities.

Practically from the moment a left majority was established on the Vancouver City Council, the city was caught up in a defensive struggle to maintain jobs and services in the public sector, in the face of provincial cutbacks. It thus was forced into the same defensive posture as the left Labour councils in Britain. It made various gestures on peace issues and prided itself on maintaining services in difficult times, but it lacked the resources for bold action on matters like housing and social services. What is more, it found itself confronted by a provincial government determined to redevelop the old railway and harbor lands exactly as it chose. Even the ALRT—the new rapid transit system—was imposed by the province, with little regard to the city's or the region's transportation plans. Again, the parallels with what happened in Britain are striking: a "left council" was pushed aside by a central government to allow "docklands" development in accordance with the demands of capital.

The New Democrats and their local allies offered nothing as challenging to capital as Labour's Alternative Economic Strategy. Instead, they argued that the government was making B.C. less attractive to capital by disrupting its social harmony. On this view, Canada's comparative advantage to the United States lay precisely in its well-developed public sector. The absence of ghettos, the universal provision of good education and health care, the relative safety of the streets, all could be seen as effects of a more advanced welfare state. Such a state depended on continuing high levels of public expenditure and employment. Thus, to "make B.C. into another Alabama" might be counter-productive, as companies that otherwise would have located in Vancouver sought more hospitable locales for their headquarters operations. Was Vancouver's vaunted quality of life—which enabled corporate executives to ski or sail within sight of their downtown offices—not also a matter of social peace and harmony? And, did this peace and harmony not depend on a strong public sector?

Mayor Mike Harcourt certainly advanced this line of thinking. The COPE aldermen on his left maintained a more militant attitude toward the government and greater distance from the business community. He, on the other hand, attempted to show that a conciliatory social democrat could bring business, labor, and government together in a cooperative relationship. It was in this context that he approached his own economic development initiatives. When he became mayor, there already were two new institutions for economic development in the city: The Vancouver Economic Advisory Commission (VEAC) and the Economic Development Office (EDO) of the City of Vancouver. His predecessor had created them both in 1978, with fairly traditional notions of business promotion in mind. Harcourt gave their activities high priority and was able, despite his NDP connections, to win sufficient high-profile business support for the Advisory Commission to make it credible. He also got backing from the Labour Council and involved academics from across the ideological spectrum in its work.

The Economic Strategy developed by the EDO and VEAC in 1981-1982, and approved by the council in 1983, focused on a set of "strategic initiatives" (Vancouver Economic Advisory Commission, 1983): namely to

(1) promote Vancouver as the key business communications center linking North America and the Pacific Rim;
(2) promote Vancouver as an emerging international financial center;
(3) promote programs to increase expenditures by tourists in Vancouver;
(4) develop a program to promote Vancouver's "invisible exports," such as business, financial, consulting, engineering, and transportation services;
(5) encourage advanced technology manufacturers;
(6) promote the development of Vancouver's health services by formulating a development plan for the city's medical precinct;
(7) support Vancouver's small-business entrepreneurs;
(8) encourage the development of light industrial space in the city.

These initiatives were to be complemented by more socially-oriented activities, including provision of more low-to-moderate-income housing near the CBD. Although item 6 implies a more positive attitude toward the public sector than the province had adopted, nothing in this "social democratic" strategy was really at odds with the ambitions of the local business community. Indeed, most of it was complementary to the initiatives of conservative federal and provincial governments.

Vancouver's development as the center for Canadian economic relations with the Asia Pacific was a concern of all three levels of government in the 1980s. The Asia Pacific Foundation (to enhance cultural ties) and the Asia Pacific Initiative (for business promotion) were launched in Vancouver by intergovernmental agreement. The two senior governments adopted complementary programs to attract immigrant entrepreneurs (especially from Hong Kong) and passed legislation to free offshore banking operations in Vancouver from certain taxes and regulations. The latter measures were intended to make Vancouver an international financial center, especially for Asian capital. When the province sold B.C. Place to Li Ka-Shing, it justified its actions as a means for cementing economic relations with Hong Kong, which was seen as the key source of new capital, new entrepreneurs, and new business connections.

Mayor Harcourt also was on the Asia Pacific bandwagon. He led various trade and economic missions to China and Southeast Asia, bringing with him not only representatives of the business community, but people from the two Vancouver universities, the Labour Council, and the civic administration (Hutton, 1985). This outreach was linked to Vancouver's sister city program, which framed various efforts to facilitate business connections with Asia. The mayor's approach was distinctive in so far as it assumed that interests other than business had a legitimate role in the development of economic policy, and that economic success did not depend on slashing public spending. These ideas were unpopular in the federal as well as the provincial capital after the Conservatives assumed office in Ottawa in 1984. No doubt Harcourt was fortunate to have begun his initiative before the climate for cooperation between business and labor began to deteriorate in the wake of the province's restraint program. As a result, he was able to get considerable support in the business community for a type of corporatist planning.

Harcourt moved into provincial politics in 1986 and assumed leadership of the NDP the following year. In that context, he has continued to appeal for a new cooperative relationship between business, government, and labor. In the city itself, however, the policy networks he developed have been allowed to atrophy. This reflects both the return to power of the conservative NPA and the easing of local anxieties about Vancouver's economic future. Under a popular new leader, Social Credit won a decisive victory in the October 1986 provincial elections, and—using much the same electoral organization—the NPA swept to victory a month later in the Vancouver municipal elections. The new mayor was Gordon Campbell, a young developer and former aide to TEAM Mayor Art Phillips, who presented himself effectively as a liberal alternative to the extremes of the left and the right.

As the Vancouver economy picked up in the late 1980s, development pressures in inner-city neighborhoods intensified and housing prices escalated. Wealthy Chinese immigrants, from Hong Kong and elsewhere, became associated in the public mind with both the rise in house prices and the construction of "block-buster" homes in established neighborhoods. As a result, concerns about growth restraint, neighborhood protection, and affordable housing came again to the fore, and Campbell and his colleagues began to edge away from pro-development rhetoric.

Campbell downplayed the role of the Economic Development Office and eventually abolished the Economic Advisory Commission. Before the Commission disappeared at the end of 1988, it produced a new draft strategy for the 1990s. The council has refused even to consider the strategy, although the analysis it adduces likely will inform some of the activities of civic agencies. The document refers to three "key perceptions and principles": (1) "Vancouver's emergence (and potential) as a business center within the urban system of the Pacific Rim"; (2) the need to preserve "the city's unique and cherished 'livability' "; and (3) "the crucial role of knowledge, entrepreneurship and innovation in effecting a successful transition to progressively more advanced stages of economic development". It is perhaps significant that the *last* of the five "strategic objectives" that emerge from these principles is to "provide the fullest possible opportunity for Vancouver's people to participate in the economic life of the city, and to increase their prosperity and socioeconomic well-being" (Vancouver Economic Advisory Commission, 1988). This clearly was a secondary concern even for a body that owed its impetus to a social democratic mayor.

TOWARD A LIVABLE CITY?

Despite the efforts of various governments with different philosophies, the Vancouver economy remains both uncomfortably dependent on the provincial resource industries and extremely vulnerable to the next downturn in the global economy. The city undoubtedly has developed as a corporate headquarters, and it has improved its links with the Asia Pacific (Davis & Hutton, 1989; Goldberg & Davis, 1988; Hutton, 1989; Hutton & Ley, 1987). However, it would be rash to claim that such changes are the consequence of public initiatives. The clearest effects of public policy have been on the quality of public services and the character of land-use in downtown Vancouver. The city itself has had far less effect in this respect than the province of British Columbia, and the province's initiatives have been almost entirely deleterious. These initiatives certainly have limited the ability of local authorities to deal with social and economic problems. Worse, the quality of life for ordinary people has deteriorated as a result of the underfunding of schools, hospitals, and other public services, and of the failure of the federal and provincial governments to develop effective housing programs for people on low incomes. This deterioration has been rationalized as an inevitable consequence of policies designed to attract outside capital and stimulate domestic investment, but neither the inevitability of the consequence nor the effectiveness of the attraction is at all evident.

Since Mike Harcourt's departure for provincial politics, what appeared as a new, social democratic style of leadership and a new type of local economic initiative has gradually disappeared. Instead, there has been a return to a more familiar style of leadership, which puts the municipality at the service of land development and inhibits public participation in strategic planning by capitalist enterprise. If this situation changes again, it will likely be as a result of a renewed sense of economic crisis—or pressure from the provincial government. If the NDP's present lead in the opinion holds up, Mike Harcourt may yet have his way as provincial premier and move the business community toward corporatist modes of economic planning. Regardless, Vancouver is bound to be at the center of any new economic strategy for the province.

REFERENCES

Allen, R. C., & Rosenbluth, G. (Eds.). (1986). *Restraining the economy: Social credit economic policies for B.C. in the eighties*. Vancouver: New Star Books for the B.C. Economic Policy Institute.

Anderson, B., & Wachtel, E. (Eds.). (1986). *The Expo story*. Vancouver: Harbour Publishing.

Blake, D. (1985). *Two political worlds: Parties and voting in British Columbia*. Vancouver: U.B.C. Press.

Davis, H. C., & Hutton, T. A. (1989). The two economies of British Columbia, *B.C. Studies, 82*, 3-15.

Evenden, L. (Ed.). (1978). Vancouver: Western metropolis. *Western geographical series* (Vol. 16). Victoria, B.C.: University of Victoria, Department of Geography.

Goldberg, M. A., & Davis, H. C. (1988, June). *Global cities and public policy: The case of Vancouver, British Columbia*. (Working Paper No. 17). Vancouver: U.B.C. School of Community and Regional Planning, Comparative Urban and Regional Studies.

Goldberg, M. A., & Mercer, J. (1986). *The myth of the North American city*. Vancouver: U.B.C. Press.

Greater Vancouver Regional District. Planning Department. (1980, September). *The livable region: From the 70s to the 80s*.

Greater Vancouver Regional District. (1988). *Achieving Vancouver's potential: An economic vision and action plan for the livable region*.

Gutstein, D. (1988). *Ka-shing in* on Privatization. *New Directions, 4*, (2), 13-16.

Gutstein, D. (1983). Vancouver. In W. Magnusson & A. Sancton (Eds.), *city politics in Canada* (pp. 189-221). Toronto: University of Toronto Press.

Gutstein, D. (1975). *Vancouver Ltd.* Toronto: James Lorimer.

Howlett, K. (1990, January 20). B.C. economy comes out of the woods. *Globe & Mail Report on Business*, p. B1.

Howlett, M., & Brownsey, K. (1988, Spring). The old reality and the new reality: Party politics and public policy in British Columbia, 1941-1987. *Studies in Political Economy: A Socialist Review, 25*, 141-176.

Hutton, T. A. (1985, October 16). *A municipal perspective on the promotion of Pacific Rim trade and investment*. Presentation to the U.B.C. Department of Geography Colloquium, Vancouver.

Hutton, T. A. (1989, August). *Vancouver as an emerging center of the Pacific Rim urban system* (Working paper). Presented to the Inaugural Pacific Rim Urban Development Council Conference, Los Angeles.

Hutton, T. A. & Ley, D. F. (1987). Location, linkages and labour: The downtown complex of corporate activities in a medium size city. *Economic Geography, 63*, 126-141.

Leo, C. (1977). *The politics of urban development: Canadian urban expressway disputes.* Toronto: Institute of Public Administration of Canada.

Ley, D. (1980). Liberal ideology and the postindustrial city. *Annals of the Association of American Geographers, 70*, 238-258.

Ley, D. (1981). Inner city revitalization in Canada: A Vancouver case study. *The Canadian Geographer, 15*, 124-148.

Ley, D. (1987). Styles of the times: Liberal and neoconservative landscapes in inner Vancouver, 1968-1986. *Journal of Historical Geography, 13*, 40-56.

Magnusson, W., Carroll, W. K., Doyle, C., Langer, M., & Walker, R. B. J. (Eds.). (1984). *The new reality: The politics of restraint in British Columbia.* Vancouver: New Star Books.

Magnusson, W., & Sancton, A. (Eds.). (1983). *City politics in Canada*. Toronto: University of Toronto Press.

Marchak, P. M. (1983). *Green gold: The forest industry in B.C.* Vancouver: U.B.C. Press.

Matas, R. (1989, June 17). Mystery, unanswered questions remain about B.C.'s land deal of the century. *Globe & Mail.*

Tennant, P. (1980). Vancouver civic politics, 1929-1980. *B.C. Studies, 46*, 3-27.

Tennant, P., & Zirnhelt, D. (1973). Metropolitan government in Vancouver: The strategy of gentle imposition. *Canadian Public Administration, 16*, 124-38.

Vancouver, City of. Planning Department. (1989, July). *Central area plan: Policy report no. 1, goals for the central area and policy directions for land use.*

Vancouver, City of. (1986). *The Vancouver Plan: The city's strategy for managing change.*

Vancouver, City of. Economic Development Office. (1985, June). *An analysis of economic structure, growth and change.*

Vancouver, City of. (1987, April). *The Expo year: 1986: Graphic Displays of selected economic indicators for the period culminating in Vancouver's year of Expo 86.*

Vancouver, City of. (1987, July). *Vancouver economic data base.*

Vancouver, City of. (1989, October). *Vancouver economic data base.*

Vancouver Economic Advisory Commission. (1980, January). *Economic goals programme for the City of Vancouver: Report with recommendations.*

Vancouver Economic Advisory Commission. (1984, April). *An economic strategy for Vancouver: Progress report no. 1.*

Vancouver Economic Advisory Commission. (1984, September). *An economic strategy for Vancouver: Progress report no. 2.*

Vancouver Economic Advisory Commission. (1982, May). *An economic strategy for Vancouver in the 1980s.*
Vancouver Economic Advisory Commission. (19183, April). *An economic strategy for Vancouver in the 1980s: Proposals for policy and implementation.*
Vancouver Economic Advisory Commission. (1988). *A strategy for Vancouver's economic development in the 1990s.*
Williams, G. (1988, Spring). On determining Canada's location within the international political economy. *Studies in Political Economy: A Socialist Review, 25*, 107-140.

Part III

Privatizing Regeneration

10

The Two Baltimores

RICHARD C. HULA

BALTIMORE: A MODEL FOR REGENERATION?

Baltimore seems an unlikely candidate to serve as a model for an urban renaissance. Generally viewed as a modest blue-collar manufacturing center, it has long been overshadowed by competing cities in the Northeast. Nevertheless, Baltimore is increasing identified as a prototype for cities seeking to generate economic and social renewal. Local boosters point to a central city that has largely rebuilt over the past 20 years and to an emerging tourist industry that has brought millions of visitors to the city. Moreover, many claim a feeling of pride on the part of city residents and a growing collective sense that it is possible to direct future development. This sense of accomplishment was strong enough to catapult William Schaefer, the mayor of Baltimore from 1971 to 1986, into the Maryland governor's office.

Recently, however, a significant debate about the Baltimore renewal effort has emerged. Critics claim that much of the redevelopment has at best had a modest impact on the bulk of the city's population. Some, in fact, argue that it has exacerbated existing racial and economic divisions within the city. Questions have also been raised about the techniques and strategies that have been used to implement the programs. This chapter begins with a review of the Baltimore renewal effort and then seeks to evaluate some of the conflicting claims that have been made about it.

THE CITY

The economic history of Baltimore is closely tied to its development as a port. In addition to having a first-rate natural harbor, the city enjoyed the advantage of being both significantly closer to Europe and further inland than other eastern ports. These locational advantages served to promote the city as a shipping and rail center. Related industries such as shipbuilding and repair also thrived. A large primary metal industry, particularly steel production, developed. Although emerging economic activity was often centered outside the physical boundaries of the city, throughout the nineteenth century Baltimore was able to capture this new economic growth by expanding its borders. However, this pattern of growth came to an end in 1918 with a final burst of annexation, which tripled the size of the city from 30 to 92 sq. mi.[1]

There has never been any question of Baltimore's secondary economic status in the region. It was (and continues to be) a city of branches and subsidiaries. Even during the 1930s and 1940s, when it could claim to be the seventh-largest manufacturing center in the country, the city had very few corporate headquarters. Today only one *Fortune* 500 corporation is headquartered in the city. This lack of local management has been associated with several problems. Major economic decisions affecting the life of the city have typically been made by actors not only unfamiliar with it, but also largely unsympathetic to its interests. In addition, economic elites residing in the city fail to identify the economic interests of the city with the businesses that they represent.

COPING WITH URBAN DECLINE

As happened in many older industrial cities in the United States, the decades following World War II were a time of demographic change and economic decline for Baltimore. The city experienced a dramatic decline in population, dropping from a maximum of 950,000 in 1950 to the current population of approximately 750,000. This is almost 100,000 fewer persons than reported by the 1930 Census. Even as the city was losing population, the Baltimore urban region continued to experience significant growth. Since 1950 the region has grown by more than one million persons. This differential is reflected in the decline of the region's population living in Baltimore City, dropping from more than 60% in 1950 to less than 35% in 1987. In addition to aggregate population loss, the demographic and economic makeup of the city also shifted significantly. In 1960 Baltimore was predominately white (65%), but nearly one-half nonwhite by 1970. Currently the city is more than 60% nonwhite.[2] Predictably, changing racial statistics were associated with an increased demand for social and welfare services.[3]

The most dramatic change in the post-World War II environment was the near collapse of Baltimore's industrial economy.[4] Total jobs in the city declined at a rate faster than that of the population. From 1970 to 1987 the city recorded a net loss of 47,000 jobs. Particularly hard hit was the manufacturing sector, which had long served as the main engine of Baltimore prosperity. The number of manufacturing establishments in the city declined from 1,738 in 1960 to 664 in 1986, a reduction of more than 60%. The proportion of the state's manufacturing located in the city declined sharply, falling from 56% in 1950 to less than 27% in 1986. This economic change caused a significant drop in manufacturing employment. Between 1970 and 1987 the city lost more than 50,000 manufacturing jobs. In addition, in excess of 14,000 infrastructure jobs in construction, communications, utilities, and transportation were lost. Such losses were only partially offset by an increase of 36,000 jobs in the service sector (Regional Planning Council, 1989).

Of specific concern was the deterioration of the port of Baltimore. The decline was due to a number of interrelated factors. The most important of these were technological. Port facilities were increasingly obsolete, with modernization proceeding very slowly due to a lack of significant private investment. Of particular importance was the failure of the port to adapt to new containerized cargo techniques that were revolutionizing shipping.[5] A second set of problems centered on efforts of shippers to reduce the power of very strong dockworker unions. Shippers began more often to turn to nonunion ports, particularly Norfolk, Virginia.

Some elements of the city's economic decline are mirrored at the regional level. For example, between 1970 and 1987, regional manufacturing employment declined approximately 30%, from 198,600 to 137,600. For the most part, however, city decline has occurred in the context of regional growth and expansion. Between 1970 and 1987 total employment increased by approximately 167,000. Almost all of this growth was centered in the service sector. If one were to exclude the economic performance of the city, regional economic growth becomes even more positive. Indeed, an index of economic health of 24 metropolitan areas, compiled by the accounting firm of Grant Thornton, consistently ranks the Baltimore metropolitan area very highly.[6] In the past decade this development has expanded beyond the traditional metropolitan region. Of particular importance is the I-95 corridor connecting Baltimore and Washington, D.C.

THE SEARCH FOR SOLUTIONS

By the late 1950s the city's economic distress was visible in the declining fortunes of the its central business district. Both commercial and retail activity were in decline. There was a clear deterioration of the downtown building stock as private investment was withheld. Indeed, no significant building had occurred in the downtown area since the 1920s (Pritchett, 1982). Demand for commercial and retail property was limited, with more than

2,000,000 sq. ft. of loft and warehouse space vacant. The city's waterfront had become a jungle of rotted piers and abandoned warehouses.[7] The future of the central city seemed bleak.

In response to these economic problems, city officials began to put together a plan for public action. Working largely independently, a number of private organizations also began to devise a public strategy to combat the urban decline. An early leader this effort was the Citizens Planning and Housing Association. The CPHA had been founded in 1941 with the goal of supporting economic and neighborhood development.[8] Active in numerous neighborhood issues, the CPHA has often provided technical advice to neighborhood groups (Froelicher, 1982; Olson, 1980).[9] An additional goal of the CPHA was the mobilization of the city's business community. In 1955 the CPHA was influential in creating the Greater Baltimore Committee. The GBC was an explicitly elitist organization comprising the chief executive officers of the city's 100 largest businesses. It was established to reverse the economic decline of the city. The Greater Baltimore Committee and the Committee for Downtown, an organization representing downtown merchants, created a private planning unit to review alternative redevelopment strategies. Initially, an effort was made to devise a comprehensive redevelopment plan for the entire downtown district. What emerged was the Charles Center redevelopment plan (Bonnell, 1982). It identified a 33-acre core in Baltimore's central business district for redevelopment. The $180-million project was to transform the area into a center of modern offices, retail establishments, and some apartments. The focused nature of the plan was justified, based on the notion that for downtown redevelopment to occur, there needed to be a single dramatic focal point to show that such development was possible.

Two critical assumptions were built into the Charles Center plan. The first identified downtown commercial interests as the key to economic revitalization. Such investment was seen as a means to integrate Baltimore's local economy into an emerging national economy, based not on manufacturing but on information and services. Such an integration was seen as having very broad distributional benefits to the entire city. A second assumption asserted the critical role of private capital. In fact, the project was conceptualized largely as a private venture to be facilitated by the local government, primarily through land acquisition and clearance. Of the estimated $180-million cost of the project, only $8.3 million was to be public funds.

The Charles Center redevelopment plan was submitted to the city in 1958. It was received with enthusiasm and quickly replaced the city's own renewal design. Lyall (1980) suggests that the city plan suffered from two major faults. It failed to gain the support of the business elite because it largely ignored the downtown. The city's plan was also greeted with hostility in a number of neighborhoods that had been targeted for freeway construction. The Charles Center project moved forward in 1959 as the Maryland legislature allowed and city voters approved a $10.5-million bond issue to support the renewal plan. That year the target area was declared an urban renewal zone for which the city received $28.3 million in federal urban renewal funds.

Actual administrative responsibility for implementing Charles Center redevelopment was put into the hands of a private corporation, the Charles Center Management Office. The use of the private management office was justified on the basis of efficiency and the need to allow sensitive negotiations with developers to proceed with confidentiality. Thus it was left to the Charles Center Management Office to acquire some 350 properties in the development area and package these into development sites. The role of the city was explicitly supportive, centered largely on the implementation of the site development plans prepared by the Charles Center Management Office.

The early success of Charles Center strongly supported the view that downtown Baltimore was in fact a viable entity. By 1961 ground was broken for the first new building, One Charles Place. A number of other commitments followed, so that by the middle of the 1960s it was clear that the project would in fact be completed. In 1964, at the request of the city's mayor, the Greater Baltimore Committee unveiled a second, more ambitious downtown renewal program. This plan proposed the revitalization of the Inner Harbor area and the construction of a municipal center complex. However, in the same year a bond issue to finance this development was defeated by city voters.

The bond issue defeat did not cause the city to give up on it hopes for Inner Harbor renovation. Rather, plans for the Inner Harbor were simply decoupled from the municipal complex, and work continued.[10] In 1965 the Charles Center Management office was incorporated as the Charles Center Inner Harbor Management Corporation and charged with the administration of both Inner Harbor and Charles Center redevelopment. Although the Inner Harbor plan was often described in terms very similar to that of Charles Center, there were important differences. It was, first of all, more ambitious, calling for a $270-million investment over 30 years. More important is the shift in relative financial commitments of the public and private sectors. The Inner Harbor plan called for a large-scale public investment. In addition to infrastructure investments such as parks, promenades, and a marina, the Inner Harbor plan called for the public construction (or at least significant public subsidy) of a World Trade Center for the Maryland Port Authority, a convention center, a new science center for the Maryland Academy of Sciences, and an aquarium. The specific goals of the plan were also different. Whereas

Charles Center focused largely on commercial development, the Inner Harbor introduced tourism as an important redevelopment goal. In contrast to Charles Center, it was assumed that large public subsidies would be required for some time to attract private investment. Simply put, the Inner Harbor venture was perceived to be a much more speculative venture. Largely unchallenged, however, were the twin assumptions of the Charles Center project, which identified the public interest with downtown development and with private implementation.

In 1968 the city declared the Inner Harbor an urban renewal area and began receiving federal funds to initiate land acquisition and clearance. Although Inner Harbor redevelopment received a strong endorsement from the city, there remained a good deal of skepticism about the plan's feasability, both in the public at large and among the city's economic elite. A measure of this skepticism is given by the failure of the CCIHMC to generate significant private investment in the area.

Although the outline of Baltimore's downtown redevelopment effort was already in place, the election of William Donald Schaefer as mayor of Baltimore in 1971 was critical to its implementation. Until he left the post in 1986 to become the governor of Maryland, Schaefer would be a tireless advocate of the renewal program. His campaign had stressed the general desirability of downtown renewal and, in particular, the need for Inner Harbor redevelopment. On assuming office, he moved quickly to invigorate the Inner Harbor plan. In addition, Schaefer initiated a wide range of development programs targeted to the industrial/manufacturing sector, middle-class housing, and neighborhood renewal.

As mayor, William Schaefer defied conventional stereotypes. He was clearly an activist mayor with a long list of policy initiatives to his credit. Schaefer's personal drive and desire to "get things done" was legendary. Nevertheless, he was a strong fiscal conservative, deeply skeptical of the ability of a public bureaucracy to implement effective public policy.[11] Thus, for much of Schaefer's term in office, the size of Baltimore government was shrinking rather than expanding. The size of the city work force declined for most of the years that Schaefer was in office. Overall, the total number of positions in the city declined more than 12% from 1977 to 1985. Total person-hours worked by city employees were reduced by 22%. The Schaefer administration adhered to a tradition of modest capital debt. Unlike many cities, Baltimore resisted the temptation to borrow on a grand scale during the 1970s and managed to maintain a reasonably strong bond rating. Overall increases in local property tax rates have been modest. Although rates have been relatively stable since 1974, they continue to be much higher than those found in other jurisdictions in the metropolitan region.[12]

REDEVELOPMENT POLICIES 1971-1986

Redevelopment initiatives implemented by the Schaefer administration can be crudely assigned to three broad categories: downtown renewal, housing, and manpower and economic development. Downtown renewal includes not only the Charles Center and the Inner Harbor but also a number of projects south and east of the downtown. Housing programs have been targeted to both low-income and affluent markets. However, Schaefer's major policy goal was to entice the affluent to live in the city. His lack of interest in low-income housing was based on his view that the economic future of the city was tied to its ability to attract higher-income residents and resist, if possible, the continued in-migration of low-income groups. Manpower and economic development included general promotional programs to attract new business as well as specific efforts in land banking and business creation. Obviously, no programs have a single well-defined goal, but rather have multiple, sometimes conflicting, aims. Moreover, individual programs interact with each other, sometimes reinforcing dominant program goals and at other times generating conflict. It should also be clear that specific programs and their associated goals are not of equal importance to political authorities. Without question, the major focus of the Schaefer administration was downtown renewal.

The apparent paradox, generated by efforts to reduce the scale of the public sector while simultaneously embarking on a series of policy initiatives to revitalize the city, can be resolved by an analysis of the privatization strategies used by the Schaefer administration to fund and implement these new policy initiatives. That is, the private sector rather than the government was used to implement social policy. The most straightforward example utilized direct subsidies to private-sector actors. Supply-side subsidies have been directed to condominium projects in the Inner Harbor. The largest supply-side project has been Coldspring new town. Coldspring was conceptualized as a $200-million project that would ultimately produce 4,000 moderate- and upper-income housing units. But while using more traditional forms of privatization, the city has come to rely most heavily on the quasi-private corporation to implement redevelopment policy. For example, the Central City Development Corporation currently implements much of the city's downtown renewal program. More specialized organizations have also emerged. Some of these deal with specific tasks such as the management of the convention center and Hyatt Hotel, the National Aquarium, and the municipal marina. Others have been given broader charges. For example, major industrial development efforts have been channeled through the Baltimore Economic Development Corporation (Berkowitz, 1984, 1987).

THE SHADOW GOVERNMENT

The complex network of institutions that have evolved around the city's redevelopment effort is sometimes referred to as Baltimore's "shadow government" (Stocker, 1987). The most prominent organization in the shadow government has been the agency charged with implementing central city renewal. There have been three such agencies: the Charles Center Management Office, the Charles Center Inner Harbor Management Corporation, and the Center City Development Corporation. The formal political status of each of these corporations is difficult to specify. Each has received significant public operating funds to design and implement redevelopment policy. In addition, each has had a good deal of influence on capital spending in the downtown area.[13] From this perspective they seem clearly to be public agencies. Nevertheless, corporation officials have consistently denied public access to the records and meetings on the grounds that they are private organizations operating under a contract with the city. In making this claim, the corporation not only secures privacy for its activities, but also avoids a number of constraints facing public officials, ranging from mandatory relocation payments for displaced individuals to competitive bidding.

The most controversial element within the shadow government was the Loan and Guarantee Program. This Loan and Guarantee program was not tied to any specific project, but rather served as an in-house development bank. It did not actively seek out loans, but rather evaluated proposals from a wide variety of public and quasi-private organizations. For example, for fiscal year 1983, the program supported 53 projects. These included 33 originated by the city's Department of Housing and Community Development, 5 from the Baltimore Economic Development Corporation, 7 from the Charles Center Inner Harbor Management Corporation, 1 with the Market Center Development Corporation, and 6 from the city's capital budget.[14] Decisions concerning specific loans were made in private, with no direct public oversight, by two trustees appointed by the mayor.[15] While loans and guarantees were subject to review and approval of the City Board of Estimate, this review was seldom very rigorous. This lack of oversight was due in part to a lack of information on the part of the regulators. For example, the trustees' staff consistently refused to publicly discuss their activities, on the principal that they were employed by a private organization. Oversight was also restricted by the strong political control maintained by Mayor Schaefer over the Board of Estimate.

The Loan and Guarantee Program was established in 1976 as a mechanism to allow the city to make an indirect loan to the developer of Coldspring (Smith, 1980d). Although the city could not legally make the loan, it could allocate funds to a trust. The trust was then free make the loan to the developer. From 1976 to 1977 the city entered into a set of contracts with the trust to administer designated city funds. With the active support of Mayor Schaefer the trustees expanded their loan portfolio to value between $100 and $200 million by the early 1980s.

One of the most remarkable features of the trustee arrangement was the difficulty in securing data on the source of loan funds and where those funds had been allocated. Some limited information on the activities of the trustees was made available after the City Council demanded a minimum annual report. To illustrate the range of activities financed by the trustees, Table 10.1 presents a listing of projects approved in 1981. While it is clear in this table that the range of projects funded by the trustees was large, downtown efforts remained the most prominent. Such projects accounted for more than $24 million of the almost $45 million committed by the trustees. The nature of projects funded by the trustees was often quite diverse, particularly those outside the downtown area. In one case, which received a good deal of notoriety in the local press, the trustees made an independent judgment that the city needed to purchase property for a landfill. Unfortunately, the land purchased was in an adjoining county, which refused to zone the property for this use.

Non-city funds were critical to the financing of trustee loans. Approximately 36% of the resources come from federal grants, most often Urban Development Action Grants and Community Development Block Grants. Approximately 27% of the funds came from city and industrial revenue bonds. Slightly more than 18% of trustee investments were from funds directly generated and controlled by the trustees. These funds have been accumulated from UDAG repayments and interest earned from short-term investments. By relying on such funds, the trustees were able to make a wide range of capital investment decisions without the City Council or voter approval that would be required in normal capital expenditures.

The trustee program generated a good deal of controversy. Some city council members and community activists attacked the institution as undemocratic and overly responsive to downtown economic interests. Some saw the trustees as an important element in the efforts of Mayor Schaefer to increase his political power base in the city. Led by Mayor Schaefer, the supporters of the trustees claimed that they were an integral part of the Baltimore redevelopment success. Indeed, a publication of the National League of Cities portrayed the trustees as an example of Baltimore's innovative approach to redevelopment, which might be utilized by other jurisdictions. With the exception of some modest reporting requirements imposed in the early 1980s, the trustee system seemed secure. However, in 1986 Mayor Schaefer announced that the trustees would be abolished, and their loan

TABLE 10.1
Summary of Trustee Investments for Fiscal 1984

Geographic Region of Project	*Trustee Contribution ($)*	*Total Value ($)*
Northwest	1,591,620	4,094,820
Northeast	1,602,450	2,767,579
West	6,198,146	22,020,076
Downtown	24,543,705	112,565,355
East	4,961,483	13,926,758
Southeast	53,800	53,800
Southwest	1,874,872	8,188,572
South	119,000	1,919,000
Citywide	3,300,000	3,300,000
Total	44,551,487	172,211,706

Downtown Project	*Type*	*Trustee Contribution ($)*	*Total Value ($)*
Omni Hotel	Commercial	4,663,155	51,163,551
Charles Plaza	Commercial	600,000	2,400,000
Ethel's Place	Commercial	745,000	1,771,500
Howard Johnson's Garage	Commercial	3,910,000	3,910,500
Lexington Market Arcade	Commercial	8,000,000	8,000,000
Lexington Market Garage	Commercial	865,500	865,500
Polytechnic Institute Conversion	Commercial	700,000	700,000
Rivoli Garage	Commercial	2,990,050	2,990,050
Beethoven Apartments	Residential	In litigation	
Chesapeake Commons	Residential	1,470,000	11,665,150
Sheridan Inner Harbor Hotel/Garage	Commercial	9,500,000	30,000,000

Source: Annual Report of the Trustee for the Loan and Guarantee Program of Baltimore City for Fiscal Year 1984

portfolio would be henceforth managed by a unit in the City Finance Department. The official reason given for ending the trustee system was that private investment was no longer difficult to attract to the city. Thus, the trustees were no longer needed. Other commentators suggest that the mayor had come to view the trustees as a political liability in his upcoming gubernatorial campaign.

The Baltimore Economic Development Corporation (BEDCO) has been another important actor in the shadow government. Created in 1976 from a merger of the City's Department of Economic Development and the Baltimore Development Corporation, BEDCO has tried to encourage industrial investment in the city. BEDCO has provided a link between the city bureaucracy and local industrial firms. It has acted as a city land bank through its administration of municipal industrial parks. BEDCO was also charged with implementing the state enterprise zone program in the city (Berkowitz, 1984).

One could cite a great number of additional examples of quasi-private corporations that were established during the Schaefer administration. While estimates vary, most observers recognize at least 30 to 40 such organizations. Obviously, some of these are quite active and spend a good deal of money; others are much less important. For some of the organizations, budget and even function data are difficult to obtain. Whereas Baltimore Aquarium Incorporated oversees an operating budget of $650,000 and supervises a multimillion-dollar capital budget, others are paper corporations created to facilitate fund transfers.

A complete description of the overall ecology of Baltimore's shadow government is even more complex than a listing of institutions would suggest. Numerous projects not directly administered by such agencies are often closely tied to the redevelopment network. An excellent example is Coldspring new town, which is linked in a number of important ways to the shadow government. The trustees were created as a means to support Coldspring. The Baltimore Development Corporation, a subsidy of BEDCO, was created as a paper corporation to permit the city to pass grant funds to Coldspring's private developer.

A key condition that permitted the development of the shadow government was the ability of the city to fund its redevelopment effort through sources other than local tax revenues. Such financial strategies not only allowed the city to avoid formal constraints imposed by the city charter, but also helped to defuse local opposition to specific projects by giving the impression that redevelopment was essentially "free" to city residents.[16] In 1987 only about 4% of capital improvement commitments were from the city's general fund. The largest sources of funds were a variety of intergovernmental transfers from federal and state government agencies.

The specific sources of nonlocal revenue have changed over time. Initially, both the Charles Center and Inner Harbor developments relied on extensive federal urban renewal funds, as well as grants from the Economic Development Administration. With the end of urban renewal, the city targeted some Community Development Block Grant funds into the renewal effort.[17] In addition, the city has been very successful in obtaining Urban Development Action Grants. Although current UDAG funds are quite limited, repayment of past UDAG loans from developers has allowed the city to acquire substantial working capital. Finally, the State of Maryland has provided a significant degree of support for land banking and industrial development.

A number of reasons have been put forward to support the creation and operation of the shadow government. Above all, it has been seen as a non-bureaucratic alternative to city government. Removing day-to-day redevelopment decisions from the public domain has been seen as both reducing the influence of politics and providing the environment needed to ensure private sector participation. For example, the CCIHMC often claimed that it was forced to operate in relative secrecy to protect sensitive negotiations with developers. Finally, Schaefer created multiple and sometimes overlapping organizations to generate interagency competition for resources. This competition was seen as a way to obtain maximum effort and efficiency in program implementation.

DOWNTOWN RENEWAL

Once initial clearance work had been completed, the Inner Harbor became increasingly popular as a public use area. In 1972 the city moved its large annual city fair to the harbor, and neighborhood fairs became common. Various entertainment and social events became institutionalized. In 1972 the USS *Constitution* was permanently moored at its new pier in the Inner Harbor, creating the harbor's first tourist attraction. Some private construction was also begun. By the late 1970s substantial public construction was underway, including a conventional center, the Maryland Science Center, and the National Aquarium.

In 1978 a public debate emerged when the Rouse Company proposed building Harborplace, a two-story, 3.1-acre, eating-and-shopping complex directly fronting the Inner Harbor. Opposition to the proposal was based on the fear that it would reduce public access and use of the harbor area. City officials, however, actively supported the proposal. The issue was put to a referendum ballot and passed by a substantial margin. Construction of the project began shortly thereafter.

The Baltimore Convention Center marked the completion of a key anchor to Inner Harbor development. The $45-million complex was opened in 1979 and completed a significant expansion program in 1986. Harborplace opened in July 1980, and exceeded the expectations of even its most enthusiastic supporters by drawing more than 18 million visitors in its first year. The completion of the National Aquarium in 1981 gave the Inner Harbor its major tourist attraction. The Aquarium attracted more than seven million visitors between 1981 and 1986, making it the single most popular tourist attraction in the state of Maryland. The city estimates that in 1984 the Aquarium alone generated $3.3 million in state taxes and more than $1.9 million in local taxes. Based largely on this success, the Aquarium is presently constructing a major addition.[18]

An important constraint on the development of a tourism industry in Baltimore was a serious lack of quality hotel rooms. Those attending meetings in the convention center, for example, were sometimes forced to find accommodations in the surrounding area, often as far away as Washington, D.C. Not surprisingly, city officials actively sought to attract a quality hotel to the Inner Harbor. However, in the late 1970s private investors remained unconvinced about the financial viability of such an investment. It became clear that for hotel construction to occur, the city would need to offer a substantial subsidy. In 1978 the city received a $10-million UDAG grant to subsidize the construction of a Hyatt Hotel.

As it became increasingly clear that Inner Harbor was in fact a commercial success, private investment began to flow rapidly to the area. Rubinstein and Gibb (1989) describe this new situation:

> With Harborplace's success assured, private developers leaped to invest in the Inner Harbor during the 1980's. Competition was keen to acquire developmental rights to the remaining parcels, which had attracted attention prior to the opening of Harborplace. Warehouses and vacant lots within two or three blocks of the Inner Harbor were converted to retail activity by developers unable to acquire land in the Inner Harbor itself [p. 16].

Given the difficulty experienced by city officials in attracting initial private investment, it is ironic that developers were particularly eager to develop hotel rooms. The number of first-class hotel rooms increased from 1,700 in 1981 to 4,500 in 1988. During the past decade seven new hotels have been built in the Inner Harbor. Beyond hotel development, there has been some further development of a tourist infrastructure, including additional restaurants, shopping facilities, and an urban amusement park.[19] There have also been several middle- and upper-income condominium projects.

In addition to Inner Harbor tourist development, significant commercial expansion has continued in the vicinity of Charles Center. At the end of 1988, nine separate office projects, with a combined total of 4 million sq. ft. of office space, were in various stages of development. This construction represents a measure of developer optimism since current rental rates suggest that the planned construction could not be absorbed into the market for several years. Developers have begun to show consideration of other areas of the city, particularly waterfront property south and east of the Inner Harbor. Work has begun in South Baltimore on a mixed-use project, which is to include six towers with 1,590 residential units, integrated within 200,000 sq. ft. of shopping and office space. East of the Inner Harbor, three projects with a total value of approximately $.5 billion are planned. These include a large shopping mall, a yachting center, and an upper-income housing development.[20] Public investment has also continued, including the construction of a new baseball stadium.[21]

HOUSING INITIATIVES

Coldspring was first proposed in 1972. Original plans called for the construction of 3,780 new housing units on 385 acres of land. Although conceived as a public/private effort, the city's role was dominant. Lyall (1980) describes the partnership as follows:

> The city would acquire the land, pay for the detailed plans, build the necessary infrastructure, and act as mortgagee in financing the final sale of the completed units. In addition the city would provide a short-term construction loan at favorable rates to the developer. The city, therefore, was the senior partner, with the junior partner, the developer, functioning primarily as builder and general contractor [pp. 47-48].

Initial funding of approximately $30 million came largely from the federal urban renewal program and the Economic Development Administration. An additional $10.3 million of the city's Community Development Block Grant allocation were targeted to Coldspring between 1974 and 1978.[22] Using the city trustees as its agent, the city provided a $5.5-million loan, to finance initial construction, from a fund created to underwrite low- and moderate-income mortgages. The city trustees later borrowed $7 million from a local bank to finance the second phase of construction.

Almost from its inception the Coldspring project has been plagued by controversy and cost overruns. The developer, citing extra expenses caused by poor work of minority contractors, demanded additional reimbursement from the city. Although the claim was disputed by some observers, the city did in fact reimburse the developer more than $1 million. In addition, almost $.5 million in loans made to the minority contractors was forgiven. Each of the expenditures was financed by the Community Development Block Grant funds.

The future of Coldspring is much in doubt. Only about 200 units have been constructed, and additional construction in the immediate future seems unlikely. In fact, the city has made it clear that it has given up on the original plan. A significant portion of the land that had been cleared for Coldspring has now been targeted to industrial development.

JOB CREATION

Increasing job opportunities for city residents is often cited as an important goal of the redevelopment effort. Indeed, downtown renewal is often justified on the basis of the economic opportunities it is thought to create. For the most part, however, job creation is seen as a natural externality, rather than an explicit requirement imposed on developers. There has been only a modest effort to make the link between development and economic opportunity more explicit. Berkowitz (1987) notes that, although the city has consistently rejected explicit exactions from developers, it does require all firms receiving any financial assistance from the city to utilize the city's Manpower Program as a first source of new hires and training.

Beyond expected externalities of downtown redevelopment efforts, the city has developed a number of programs that are explicitly targeted to either retain or expand the industrial base of the city. As noted, the Baltimore Economic Development Corporation (BEDCO) has been given primary responsibility in the implementation of these efforts. Acting as an agent for the city, the BEDCO has aggregated more than 500 acres in a number of industrial parks across the city. The BEDCO improves or installs appropriate infrastructure, such as streets and utilities, within the parks and then markets sites to industrial firms. The BEDCO has also been active in providing improved streets, utilities, and parking to a number of older industrial districts within the city (Berkowitz, 1987). In addition to the retention and attraction of established industry, the city has also begun some efforts to stimulate new industry. Initial efforts centered on the development of two incubator programs. The Raleigh Industrial Center, opened in 1980, was created in an eight-story, 400,000-sq.-ft. vacant manufacturing plant. A second incubator building is the Control Data Corporations's Business and Technology Center, located in one of the city's industrial parks.

THE IMPACT OF REGENERATION

Many observers see the Baltimore redevelopment program as a model for other cities. Writing in *Fortune* magazine, Gurney Breckenfeld (1977) presented the popular view:

> By almost every standard measure of trouble, Baltimore should be firmly trapped in the vortex of urban decay. It is an old, conservative, blue-collar industry place. . . . Yet Baltimore, which H. L. Mencken called the "ruins of a

TABLE 10.2
Projections Versus Outcomes for Inner Harbor Redevelopment, 1985

	Projected	*Actual*
Office Space	1.6 million sq. feet	4 million sq. feet
Retail/Tourist Space	150,000 sq. feet	1 million sq. feet
Hotel Rooms	200/300	3,500
Parking Spaces	3,000	8,000
Investment	$270 million	$2 Billion
	(50% Private)	(75% private)
Housing	2,700 Units	1,200 Units

SOURCE: Adapted from Samuels (1989)

> once great medieval city" is making an extraordinary comeback, both physically and—more important—psychologically [p. 196].

For Breckenfeld the source of Baltimore's success is clear:

> The critical ingredient of that revival is two decades of intelligent teamwork between local officials and private business leaders. Their strategy, established at the outset, has been to convert the heart of the city into a culturally rich, architecturally exciting magnet where both affluent and middle class families will choose to work, shop and live [p. 196].

Certainly city officials can point to convincing evidence of a physical renewal, much of it the result of private investment. Table 10.2 outlines the goals and results of Charles Center-Inner Harbor redevelopment. With the single exception of housing production, the city dramatically exceeded all formal goals. Moreover, the city has, in defiance of conventional wisdom, transformed itself into a tourist center.

Some observers have raised questions about the quality of the physical development. They note that earlier design standards seem to have been weakened during the 1980s. Gunts (1989) argues that architectural standards were relaxed as the city sought to maintain a steady stream of private sector investment. Design issues have become particularly important as developers question height and other limits, not only in the Inner Harbor area but also in the neighborhoods that surround it.

A more difficult evaluation question is the distribution of benefits generated from the renewal that has occurred. This is particularly true for employment. While there is no dispute that downtown renewal has created jobs, there is most certainly controversy over their number and quality. Admirers and detractors of redevelopment have made sweeping contradictory claims. Unfortunately, there are no reliable estimates of employment gains that can be directly attributed to the redevelopment process. In an effort to detect the aggregate impact of redevelopment on the city's economy, Table 10.3 presents an overview of the employment sectors in the city in 1970, 1980, 1987, and estimates for 1990. This table documents the general decline that has occurred in the local economy of the past two decades. Note, however, that recent estimates suggest that this decline may have ended. It is estimated that total jobs will actually show a modest increase (+1%) between 1987 and 1990. At this general level it is impossible to link redevelopment with this increase. Rather, one needs to consider those job sectors most closely associated with the renewal program. Three broad employment goals have dominated development efforts: the retention of manufacturing jobs, the expansion of a service economy, and the creation of a tourist/convention industry. Table 10.3 makes it clear that efforts to retain the manufacturing base of the city have had little success. Sharp reductions in the manufacturing, infrastructure, and trade sectors continued throughout the period, leaving little doubt that deindustrialization will continue into the 1990s.[23]

Table 10.3 does show a steady expansion in service employment. The medical sector is clearly its fastest-growing element, with more than 15,000 new jobs created between 1970 and 1987. The financial-service sector contributed more than 6,000 new jobs, and social and legal service each saw a growth of approximately 4,000 jobs. These data are consistent with the city's emphasis on service-sector employment. However, it is difficult to construct a convincing argument linking the large growth in the medical sector with local economic development policy. Public investment to date in this area has been quite modest. A claim that downtown renewal played an important role in generating the expansion of other service sectors, which contributed significant numbers of new jobs, is certainly not proven, but is a reasonable hypothesis.

Perhaps the most interesting finding of Table 10.3 is the lack of evidence for a major impact of the tourist/convention industry on the city's job market. There seems little support for widely quoted claims that the convention industry has created more than 20,000 jobs (*Baltimore Sun,* 1989). For example, although the hotel sector did double in size between 1979 and 1987, this expansion provided a total of less than 2,000 new jobs. Retail trade (which includes restaurant employment) has continued a dramatic decline in spite of what many claim is an expanded retail base in the Inner Harbor.[24]

Even more problematic than aggregate job creation is the issue of who actually benefits from jobs that are created. Levine (1987a, 1987b) forcefully makes the case that the physical redevelopment of the downtown area has

TABLE 10.3
Overview of Employment in Baltimore 1970-1990[a]

	1970	*1980*	*1987*	*1990*	*70/87 Change*	*87/90 Change*	*70/87 %Change*	*87/90 %Change*
Total	473.7	452.9	426.6	430.3	–47.1	+3.7	–10	+1
Infrastructure	60.1	52.1	43.1	45.9	–17.0	+2.8	–28	+6
Farming	0.5	0.5	0.4	0.4	–0.1	0	–20	0
Construction	18.3	14.4	15.1	14.0	–3.2	–1.1	–17	–7
TCU[b]	41.3	37.2	27.6	31.5	–13.7	+3.9	–33	+14
Manufacturing	99.6	68.7	49.1	47.8	–50.5	–1.3	–51	–3
Durables	45.3	31.3	17.9	17.5	–27.4	–0.4	–60	–2
Nondurables	54.3	37.4	31.2	30.3	–23.1	–0.9	–43	–3
Trade	96.8	81.4	83.0	83.1	–13.8	0.1	–14	0
Wholesale	27.3	24.9	25.8	25.7	–1.5	–0.1	–5	0
Retail	69.5	56.5	57.2	57.4	–12.3	+0.2	–18	0
Services	130.6	148.4	167.2	168.4	+36.6	+1.2	+28	+1
Financial	31.4	34.9	37.5	36.4	+6.1	–1.1	+19	–3
Medical	25.1	35.2	40.6	41.3	+15.5	+0.7	+62	+2
Business	21.1	19.3	22.4	23.6	+1.3	+1.2	+6	+5
Hotels	1.8	1.8	3.7	NR[c]	+1.9		+106	
Personal	6.9	4.4	4.9	NR	–2.0		–29	
Auto Repair	3.5	3.2	3.8	NR	+0.3		+9	
Misc. Repair	1.6	1.3	1.3	NR	–0.3		–19	
Motion Pictures	1.0	0.9	0.8	NR	–0.2		–20	
Recreation	2.1	1.5	2.4	NR	+0.3		+14	
Legal	2.6	4.5	6.8	NR	+4.2		+162	
Pvt. Education	11.8	12.6	13.8	NR	+2.0		+17	
Social	2.7	6.4	7.1	NR	+4.4		+163	
Membership	9.0	11.1	9.3	NR	+0.3		+3	
Pvt. Household	7.2	7.3	7.1	NR	–0.1		–1	
Miscellaneous	2.8	4.0	5.7	NR	+2.9		+104	
Other[d]	53.0	59.0	66.7	67.1	+13.7	+0.4	+27	+1
Government	86.6	102.3	84.2	85.1	–2.4	+.9	–3	+1
Federal Civilian	16.2	15.1	13.0	13.3	–3.2	.3	–20	2
Federal Military	7.8	4.8	3.8	3.9	–4.0	.1	–51	3
State	16.9	32.4	33.4	NR	+16.5		+98	
Local	45.7	50.0	34.0	NR	–11.7		–26	
State and Local	62.6	82.4	67.4	67.9	+4.8	+.5	+8	1

a. Reported in thousands of jobs.
b. Transportation, Utilities and Communication
c. NR indicates category not reported in 1990 estimates
d. Other Category aggregated from hotel to miscellaneous category

SOURCE: 1970-1987: Regional Planning Council (1988b) *Economic Indicators for the Baltimore Region.* For 1990 estimates: Regional Planning Council (1988a) *Employment Trends in the Baltimore Region.*

had remarkably little positive impact on the overall economy. He cites economic data showing that Baltimore's economy has declined both in absolute terms and relative to other similar sized cities through the 1970s and 1980s. For example, a Congressional Joint Economic Committee study, based on 1980 census indicators such as poverty rates, growth in per capita income, and unemployment, ranked Baltimore the fifth-neediest city in the country. The city was unranked in 1970. Poverty rates and levels of unemployment in Baltimore's poorest neighborhoods have increased, rather than declined, during the past two decades. There is little dispute with the general point that the overall economy of the city has continued to deteriorate. The city's current mayor, Kurt Schmoke, summed up this view by noting: "Baltimore is prettier but poorer in 1989 than 1979" (Topscott, 1990).

Public debate about the distributive impact of redevelopment was stimulated by the 1986 publication of *Baltimore 2000* (Szanton, 1986). This report, prepared for the Morris Goldseeker Foundation, offers a review of the current status and likely future of the city. It takes a decidedly pessimistic view of current trends, arguing that the immediate future of Baltimore is one of decline, not renaissance. This process of decline is seen as a pattern of uneven development that would, in effect, create two Baltimores:

> Baltimore may complete a pattern, already visible, of a "double-doughnut" of concentric rings. The center would contain a business, cultural and entertainment center that remained strong because it served the whole metropolitan area, and attractive housing for the well-to-do. The center would be ringed by the decaying and much more populous neighborhoods of poor and dependent, very largely black. These in turn would be surrounded by middle- and upper-income suburbs, very largely white [Szanton, 1986, p. 21].

Baltimore 2000 denied any automatic link between downtown prosperity and neighborhood well-being. This conclusion is consistent with other work suggesting that, even where downtown renewal did create jobs, they were relatively low paying and had little potential for advancement. The limited number of more attractive jobs often went to non-city residents. The report called for a dramatic shift of priorities to neighborhood and community issues, stressing the importance of infrastructure investment, particularly in education.

The vision of "two Baltimores" is reinforced in both the employment and the housing markets. The city is increasingly populated by citizens who are unprepared to compete in a technological/service economy. For example, Levine (1989) argues that between 1970 and 1984 the number of jobs requiring less than a high school education were reduced by 25%. Entry-level jobs requiring at least two years of college education increased 56%. Szanton (1986) notes that more than 50% of the jobs located in the city are filled by non-city residents, and that without a dramatic increase in the skill and education level of city residents, this proportion will continue to increase. A similar dichotomy exists within the housing market. At present there are more than 5,000 vacant units in the city, waiting for either demolition or renovation. Many of these units can be purchased for less than $30,000. Yet the average price of a new home in Baltimore is more than $200,000 (*Baltimore Sun,* 1989). This apparent anomaly is explained by the fact that practically all new construction in the city is waterfront development targeted to the affluent.

A key issue in Baltimore centers on the political process that has created and implemented the redevelopment effort. While there is certainly room for genuine debate over the independence of Baltimore's network of redevelopment agencies and corporations, it is difficult to dispute that a great number of critical economic decisions have been made by private organizations with only very limited public input. Critics have made the explicit argument that the city's development policies have had the effect of significantly reducing the autonomy of local political officials. If one is committed to private sector investment, one is inevitably constrained by private sector values. Consider the reaction of the City Council to two recent requests by developers. The first was a proposal by the IBM Corporation to add an additional 26 stories to its Inner Harbor office building. The second was a request by the Rouse Corporation to raze what was widely regarded as a historical building, and operate a temporary parking lot beyond the normal period allowed for such facilities by the city. Both requests were in clear violation of city policy, but were reluctantly approved. The majority of the council cited a continuing need to be responsive to private developer needs. Even with a history of significant redevelopment, and an apparent high demand for additional development opportunities, city officials felt unable to impose even modest restrictions on private investment.[25]

A CHANGING FUTURE?

In 1986 William Schaefer resigned as mayor of Baltimore to become the governor of Maryland. In 1987 Kurt Schmoke was elected the city's first black mayor. There has certainly been no dramatic break with past development policy. Indeed, the Schmoke administration has pledged not to impose new requirements on existing development projects. It has enthusiastically supported a number of large, upscale development projects, including further development of the Inner Harbor. Current plans call for the building of a marine center to include a 22-story research center, a maritime museum, a marine mammal pavilion, a public plaza, and a conference center. Nevertheless, there are some clear signs that local political leaders, both in and out of the Schmoke administration, are beginning to reassess important elements of Baltimore's redevelopment policy. This reassessment has, to some degree, been forced on the city since federal funds, which had provided much of Baltimore's public investment, are no longer available. Nor is there much likelihood that federal revenue will soon be replaced by other sources. The state, while expressing abstract concern for its plight, seems unwilling to significantly increase expenditures in the city. The relatively high tax rate in the city makes it even more unlikely that local revenues can be raised significantly.

A good example of fiscally driven policy change can be seen in recent strategies adopted by the Baltimore Economic Development Corporation. The BEDCO has reduced its traditional emphasis on land banking to focus on supporting indigenous economic firms. This support is seen as reinforcing, rather than replacing, private sector investment. The corporation currently is playing this role in three large development efforts. The first is the Johns Hopkins Bayview Research Center, a projected 1.5-million-sq.-ft. research center. The second is Port Covington Commons Waterfront Business Park, a 150-acre development to be anchored by a new $90-million printing plant for the *Baltimore Sun.* Plans call for the construction of almost 2.5 million sq. ft. of office and commercial space, including 10 office buildings, a 450-room hotel, a number of retail stores, and a large parking garage. Finally, the BEDCO is supporting Inner Harbor East, a $300-million largely residential development.

In addition to reacting to fiscal constraints, current leadership in the city is reviewing the process by which redevelopment policy is made. Mayor Schmoke has begun to reduce the complexity and independence of the many quasi-private corporations created during the Schaefer years. One important administrative reform merged the Charles Center-Inner Harbor Management Corporation and the Charles Center Management Corporation into a single downtown redevelopment agency, the Center City Development Corporation. The City Council has also begun to assert greater authority in the redevelopment policy. In 1989 the council angrily voted to cut the budget of the CCIHMC by 50%. The agency was instructed to request its remaining budget in six months. Approval of these additional funds was made contingent on a more open policy-making process in the agency. The council was particularly upset by the role played by the CCIHMC in pressuring it to approve the expansion of IBM's Inner Harbor building.

Modest substantive changes are also occurring. Schmoke has offered a guarded endorsement of *Baltimore 2000* and has, at least in symbolic terms, placed a greater emphasis on neighborhood redevelopment. To implement these goals some effort has been given to revitalize the local planning process. Specifically, the city moved to reduce the pace of waterfront development as it reviewed objections raised by neighborhood organizations to high-density projects. In a further response to citizen concerns, a new master plan was written for East Baltimore. A similar effort was planned for South Baltimore. In both neighborhoods the city has attempted to facilitate citizen comment and review of proposed redevelopment projects. It has made a commitment to aggressively protect and support those remaining commercial and industrial firms threatened by redevelopment. The city is also in the process of revising its downtown master plan in an effort to increase public control of future physical development there. Whether such changes merely represent a pause, or signal a fundamental shift, in policy defines Baltimore's key political issue for the next decade.

NOTES

1. The annexation of 1918 represents something of a high-water mark for the city. The annexation seemed to assure long-term economic and demographic growth. The 1920 Census reported the city had reached a population of 729,000. It was widely expected to reach one million by 1930. It was a goal the city would never reach. See Olson (1980).

2. "Nonwhite" is a category used by the 1960 Census. Later Census data indicates that a large majority of the Baltimore's nonwhite community is Black. Baltimore has a relatively small Hispanic population.

3. The increase in poverty rate is a function of the larger proportion of minorities who are poor. However, over the past decade the poverty rate for minorities in the city has remained fairly constant.

4. For an overall view of economic change in the Baltimore metropolitan area see Regional Planning Council (1988a, 1988b, 1988c).

5. The container technology demands a system of a few major ports, which then send containers on older, slower ships to older ports. By the earlier 1970s it was clear that, in the future, Baltimore would be considered a secondary port. See Olson (1980).

6. The index has been reported quarterly since 1985. It is composed of five indicators: average weekly hours of manufacturing employment, total nonagricultural employment, retail sales, business starts, business failures, and the money supply. See *Daily Record* (1989). It should be noted that although the regional economy is clearly stronger than that of the city, it is not that impressive by national standards or by those of neighboring metropolitan areas. For example, of the closest five metropolitan areas, only Philadelphia has created jobs at a slower rate (Szanton, 1986, pp. 3-4).

7. There is ample evidence that a similar process of disinvestment and physical decline was occurring in many of Baltimore's residential neighborhoods. Real property values declined in most areas of the city between 1950 and 1958. (See Levin, 1980.)

8. There is a long tradition of civic organizations framing, evaluating, and sometimes implementing local economic development policy in Baltimore. For example, in the 1930s the Industrial Factory Site Commission operated to identify manufacturing sites for businesses

wishing to locate in the region. In 1937 the Citizens Housing Association was formed to monitor the city's implementation of legislation designed to improve slum housing. In 1941 this group became the Citizens Planning and Housing Association. The CPHA expanded its scope beyond narrow issues of low-income housing to more general questions of community and economic development.

9. The CPHA has also played an important recruitment and socialization function for the formal leadership of the city. A remarkable number of elected and appointed positions are staffed with individuals who were active in the CPHA. Froelicher (1982) provides a number of examples.

10. From 1964 to 1968 city voters did approve a series of bond proposals to support inner harbor projects.

11. It should be noted that such mayoral styles are increasingly common in U.S. cities. Clark and his associate have described such a type as New Fiscal Populism. See Clark and Ferguson (1983).

12. For example, current property tax rates in the City of Baltimore are approximately twice that of Baltimore County.

13. The limits of this influence are a subject of some debate. Levine (1987a, 1987b) argues that the city has essentially delegated its decision authority to the corporation. Berkowitz (1987) strongly rejects this view and claims that the city has taken an assertive leadership role over the corporation. Whatever position one takes, however, it is clear that the corporation's influence on capital expenditures is far from negligible.

14. These data tend to underestimate the importance of projects originated by the Charles Center Inner Harbor Management Corporation. Although the CCIHMC had fewer projects funded than other submitting agencies, the project approved were typically much larger.

15. The Loan and Guarantee Program was commonly referred to as simply "the trustees" in recognition of the importance of the two men running the program. See Smith (1980a, 1980b, 1980c, 1986). The Schaefer administration has established a new kind of political organization, a corporate machine in which the trustees have become distributors of patronage in the form of low interest loans to Schaefer supporters and politically important community corporations (Levine, 1986).

16. Not surprisingly, the use of nonlocal "off-budget" resources makes it difficult to accurately track expenditures or even occasionally substantive policy.

17. Because of the targeting requirements of the CDBG, the city was unable to devote a large portion of block grant funds to downtown redevelopment. Rather, these funds were allocated to neighborhood projects and, for a time, Coldspring. The use of funds forced the city to increase the emphasis on the low- and moderate-income component of the Coldspring project. Some estimates claim that the city targeted almost 25% of the its CDBG funds to Coldspring in the early years of the program.

18. Funding for this addition follows a familiar pattern. Approximately half the funds have been obtained from private funds. The state has contributed funds. The city has contributed the land and a variety of supplementary supports. Note that once again the city's payment (which in dollar terms is estimated to be equal that of the state) is relatively hidden.

19. Not all of these projects have been successful. The amusement park has been one of the more conspicuous failures in the Inner Harbor.

20. The shopping mall is being partially funded by an interest-free loan derived from a UDAG grant. In return for this support, the city will receive an equity position in the mall.

21. Funding has been approved for a second stadium, adjacent to the baseball stadium, for football if the city is able to attract a National Football League team.

22. This represents approximately 15% of the city's total allocation for this period.

23. Within these general trends, however, there are some surprises. Note, for example the loss of more than 3,000 construction jobs. The increase in retail trade (which includes restaurants) during the 1980s may likely be associated with redevelopment.

24. Related to the issue of benefits is the question of costs associated with the development. Although it is clear that convention and tourist trade generate demands for significant city services, such costs are seldom presented in conjunction with estimates of revenue generated by the same trade. See Judd (1988, pp. 386-400).

25. The debate surrounding the IBM Building was of particular importance in that it called into question a number of long-held assumptions about the physical design of the Inner Harbor renewal. Specifically, it has forced a re-examination of the principle that development near the harbor should be of limited height to assure visual access.

REFERENCES

Berkowitz, B. L. (1987). Rejoinder to downtown redevelopment: A critical appraisal of the Baltimore renaissance. *Journal of Urban Affairs, 9* (2), 125-132.

Berkowitz, B. (1984). Economic development really works: Baltimore, Maryland. In R. D. Bingham & J. P. Blair (Eds.). *Urban affairs annual reviews: Vol. 27. Urban Economic Development.* Beverly Hills, CA: Sage.

Bonnell, B. (1982). Charles Center-Inner Harbor Management Inc. In L. H. Nast, L. N. Krause, & R. C. Monk (Eds.), *Baltimore: A living renaissance.* Baltimore: Historic Baltimore Society.

Breckenfeld, G. (1977, March). It's up to the cities to save themselves. *Fortune*, pp. 196-206.

Charles Center Inner Harbor Management Corporation. (1989). *Charles Center-Inner Harbor Projects.* (mimeo).

City of Baltimore. (1974-1986). *Summary of adopted budget.*

City of Baltimore. (1984). *Annual report of the trustees for the loan and guarantee program of Baltimore City for FY 1984.*

City of Baltimore. (1983). *Annual report of the trustees for the loan and guarantee program of Baltimore City for FY 1983.*

Clark, T., & Ferguson, L. (1983). *City money.* New York: Columbia University Press.

Convention center expansion. (1989, May 18). *Baltimore Sun*, p. A18.

Froelicher, F. M. (1982). CPHA's impact 1951-69: A personal appraisal. In L. H. Nast, L. N. Krause, & R. C. Monk (Eds.), *Baltimore: A living renaissance.* Baltimore: Historic Baltimore Society.

Gunts, E. (1989, April 2). Scarlet place and the master plan. *Baltimore Sun*, p. L1.

Judd, D. (1988). *The politics of American cities.* Boston: Scott, Foresman and Company.

Index area economy grows. (1989, March 30). *Daily Record*, p. 3.

Levin, M. (1980). *The viable city: Baltimore in the 80's.* Baltimore: University of Maryland at Baltimore.

Levine, M. V. (1986, March 7). The case for two Baltimores. *Baltimore Sun*, p. B1.

Levine, M. V. (1987a). Downtown redevelopment as an urban growth strategy: A critical appraisal of the Baltimore renaissance. *Journal of Urban Affairs, 9* (2), 103-123.

Levine, M. V. (1987b). Response to Berkowitz, economic development in Baltimore: Some additional perspectives. *Journal of Urban Affairs, 9* (2), 133-138.

Levine, M. V. (1988, January 10). Economic development to help the underclass. *Baltimore Sun.*

Levine, M. V. (1989). Urban redevelopment in a global economy: the cases of Montreal and Baltimore. In R. V. Knight & G. Gappert (Eds.), *Cities in a global society.* Newbury Park, CA: Sage.

Lyall, K. (1980). A bicycle built-for-two: Public private partnership in Baltimore. In R. S. Fosler & R. A. Berger (Eds.). *Public-private partnership in American cities.* Lexington, MA: D.C. Heath and Company.
Maryland Department of Economic and Employment Development. (1988). *Maryland statistical abstract.*
Olson, S. H. (1980). *Baltimore: The building of an American city.* Baltimore: The Johns Hopkins University Press.
Pritchett, M. (1982). Charles Center. In L. H. Nast, L. N. Krause, & R. C. Monk (Eds.). *Baltimore: A living renaissance.* Baltimore: Historic Baltimore Society.
Regional Planning Council. (1989). *Annual development report: 1988.* Baltimore: Regional Planning Council.
Regional Planning Council. (1988a). *Employment trends in the Baltimore region 1970-1990.* Baltimore: Regional Planning Council.
Regional Planning Council. (1988b). *Economic indicators for the Baltimore region.* Baltimore: Regional Planning Council.
Rubinstein, J. M., & Gibb,A. M. (1989). *Comparing urban renewal programs in Baltimore, Maryland and Glasgow, Scotland.* Paper presented at the 1989 meetings of the Urban Affairs Association.
Samuel, P. (1989, March 16). Core city problems examined. *Daily Record*, p. 1.
Smith, C. F. (1980a, April 13). Two trustees and 100 million dollar "bank" skirt restrictions of city government. *Baltimore Sun*, p. A1.
Smith, C. F. (1980b, April 13). City trustees not hampered by checks ruling other officials. *Baltimore Sun*, p. A8.
Smith, C. F. (1980c, April 13). Schaefer defends use of the trustee system. *Baltimore Sun*, p. A1.
Smith, C. F. (1980d, April 16). City rescue of Coldspring apparently broke U.S. regulations. *Baltimore Sun*, p. 11.
Smith, C. F. (1986, March 16). Trustee process was controversial in small circle. *Baltimore Sun*, p. A1.
Stocker, R. P. (1987). Baltimore: The self-evaluating city. In C. N. Stone & H. T. Sanders, *The politics of urban development.* Lawrence, KS: University of Kansas Press.
Szanton, P. L. (1986). *Baltimore 2000: A choice of futures.* Baltimore: Morris Goldseeker Foundation.
Topscott, R. (1990, January 1). Baltimore asking—nicely—for more state aid. *Washington Post*, p. D1.

11

A "Better Business Climate" in Houston

ROBERT E. PARKER
JOE R. FEAGIN

ENTREPRENEURIALISM IN HOUSTON

In many other cities entrepreneurial elites have created new political coalitions, developed public-private partnerships, and invested in infrastructure to secure expanded private investment. We have demonstrated in a recent book (Feagin & Parker, 1990) that city officials across the nation use many incentives to entice capitalist investors, including tax abatements, direct subsidies, and extensive infrastructure investments. In the case of Sunbelt cities like Houston, an aggressive, entrepreneurial elite is not new, but has, in fact, run the city from the very beginning. Houston's business elites have been particularly successful in creating *private-public partnerships* in which governments are little more than the sycophants and servants of business. Urban "Reaganomics" and "urban entrepreneurialism" were pioneered in Houston long before they were given these names.

Conservative think tanks, such as the Adam Smith Institute in Great Britain, utilized the Houston case as a primary example of the economic prosperity that comes from an unrestrained free enterprise ("supply-side") approach to economic investment and development. In an Adam Smith Institute book, *Town and Country Chaos*, Robert Jones extolled the virtues of Houston's entrepreneurial elite approach and its lack of governmental planning and public sector intervention in the local economy. He used the Houston case as a model for British cities (Jones, 1982, pp. 23-25). From the 1970s to the 1980s leading political figures in both the United States and European nations cited the unplanned free enterprise economy and entrepreneurial philosophy in Houston as the reason for the city's remarkable record of material progress.

But as we will demonstrate, Houston's business elite was ill-prepared to deal with the economic decline of their free enterprise model in the mid- and late-1980s; their belated response to seeking economic diversification for the city was relatively impotent, given the fundamentals of the internationally focused, oil-rooted metropolitan economy.

HOUSTON'S ECONOMIC STRUCTURE

Between the late 1930s and the early 1980s Houston experienced a robust economic expansion. From 1962 to 1980 the employment growth rate in the Houston area averaged 5.4% a year, compared to 2.8% for the U.S. as a whole. Between 1974 and 1981, nearly 700,000 jobs were added to the area's economy as nonfarm payroll employment rose 75%. During the same period, per capita income among Houston's workers was increasing faster than the national average, advancing 160%, compared to 108% for all U.S. workers in metropolitan areas (Greater Houston Chamber of Commerce, 1989a).

Houston has long been at the center of a global oil industry. In the early 1980s, just before the collapse in oil prices, Texas led the nation in the number of drilling rigs, oil and gas production, refinery operations, and gas reserves. Between 1973 and the early 1980s the value of oil and gas in Texas fields grew in excess of 530%, despite a 28% decline in the amount of oil actually produced (Feagin, 1988, p. 81). In the 1980s approximately 35% of the jobs in Houston were directly tied to the oil-gas industry, with another 20% heavily dependent on it. As the decade of the 1960s came to a close, multinational companies consolidated operations in Houston. Consolidation continued through the 1980s. The former president of the Houston Economic Development Council, Lee Hogan, has recently emphasized the lesser, yet continuing importance of the energy sector to the Houston economy: "Houston is now a bigger part of the energy industry . . . and the energy industry is a smaller part of Houston. Those two things are both very desirable. In our economic development, we do not want to give up the pre-eminent position we hold in that industry" (Nichols, 1988).

The Port of Houston has been a central component in the city's economic development. The value of foreign trade through the Port of Houston increased from $2.4 billion in 1970 to $23 billion in 1980. Houston's major trading partners in the early 1980s were Mexico, Saudi Arabia, and Japan. Important exports during this time included construction and mining equipment, oilfield machinery, unmilled wheat and corn, organic chemicals, and nonelectrical machinery. Houston has generally ranked as the second or third port in the U.S. in terms of total cargo tonnage and foreign trade.

The Johnson Space Center of the National Aeronautics and Space Administration (NASA) is the major space-defense complex in Houston. In the early 1980s, it employed 10,000 workers, two-thirds of whom were nongovernmental employees. The federally subsidized facility provides a substantial boost to the area's economy and is the single major source of research and development funding. In 1977, when Houston was the nation's fifth-largest city, it only ranked tenth among metropolitan areas in federal research and development expenditures. Amounts received by cities such as Los Angeles ($3.2 billion), Washington ($1.6 billion), and San Francisco ($1.4 billion) greatly outdistanced that provided to Houston ($393 million). But as we will note later, Houston began to receive a greater infusion of federal expenditures for space programs later in the 1980s, and this money has created a significant ripple effect in the local economy as companies capitalize on commercial applications.

As of the early 1980s, the impact of computers on Houston's economy had been rather limited. In the early 1980s only three companies were manufacturing a sizable number of computers: Compaq, Texas Instruments, and Zaisan, Inc. On the other hand, medical technology and its applications have had a substantial influence. In brief, Houston has become a national medical center. By the early 1980s, the Texas Medical Center consumed 235 acres and included 29 hospitals and research institutions. Overall, the economic impact has been placed at $1.5 billion. But the inability of the Texas Medical Center to generate corporate spin-offs has been disappointing.

HOUSTON'S BRAND OF POLITICS

Since Houston's founding in the 1830s, the power structure of the city has been dominated by a succession of business leaders, including cotton merchants, real estate developers, oil and railroad entrepreneurs, lawyers, and high-level corporate executives. In 1836 the Allen brothers bought 6,000 acres of swamp land in southeast Texas for a little more than one dollar an acre and laid out their projected city in a gridiron pattern, a layout considered profitable for speculative land sales. The most powerful and cohesive business leadership clique in the city's history emerged a century later, in the 1930s. Known as the "Suite 8F crowd," this entrepreneurial group was integrally involved in Houston's development until its power began to wane in the 1970s. This "Suite 8F crowd" was extraordinarily influential in shaping Houston's economic development; many of these entrepreneurs were rooted in the major local corporations in oil and gas, banking, and real estate. They were critical in creating a modern weak-government, unplanned, free enterprise city.

Since the mid-1970s, the Houston Chamber of Commerce has been at the center of the city's business leadership. Renamed the Greater Houston Chamber of Commerce in 1987, the organization has included executives of most of Houston's major corporations. As such, it is a much larger and more diverse entity than that represented by the "Suite 8F crowd." And its ability to influence the pace of development and the general direction of the city has been much less influential than the "Suite 8F crowd" because of the existence of an oil-rooted economy and the more complex world economy in which Houston's oil dynamic is now centrally located.

From the beginning, business leaders have directly or indirectly dominated mayoral and city council positions in Houston. Only rarely, such as in the 1880s, has the business dominance of local politics been challenged by members of the working class. In contrast to other U.S. cities, Houston has rarely experienced grass-roots input into local political decisions. There have been no immigrant political machines. Virtually without interruption, the local business elites have been able to exercise dominance over governmental decisions, including key domains such as planning and zoning. Since its founding, the city's business leadership has been able to implement its basic approach to government. This policy approach features three key ingredients: control over political officials, the use of public expenditures for business goals, and limiting the scope of government regulation.

For decades City Hall has been organized around a strong mayor; all city department heads, except the city controller, report directly to and are appointed by the mayor. The centrality of the mayor's power in the structure of Houston's government has generally facilitated the dominance of the local business leadership. The mayor often meets with business leaders concerning major projects before meeting with department heads and the city council. One result is that major business projects generally receive priority over social programs. All of Houston's mayors in the twentieth century have been very business-oriented. A few, though, such as Roy and Fred Hofheinz and Kathy Whitmire, have emerged from the moderate wing of the city's business community. These mayors, particularly in recent years, have been able to somewhat successfully confront the conservative majority on social issues, such as the rights of minority group citizens. Yet even these leaders have retained a strong allegiance to most business goals, including economic growth.

City planning is one arena in which the connection between the business elites and traditional political relationships can be highlighted. The business elites have sharply restricted the role of the city's planners. According to the chief of Houston's comprehensive planning division in the City Planning Department, large private projects are planned by developers with little or no input from city departments and agencies. As he put it, "In Houston the project just happens" (Feagin, 1988, p. 182). From 1964 to 1983, Houston's City Planning Department was largely relegated to researching, record keeping, and plat checking—in short, to documenting the city's growth. Because of its restricted role, other state agencies and even private companies have had to pick up the slack in Houston's planning process. For example, the Harris County Flood Control District and the Texas Highway Department have taken charge of long-range planning and infrastructural construction efforts for the city. Despite recent and successful attempts to expand planning in the city government, the planning changes still leave Houston far behind other cities in the degree of governmental intervention in urban development.

The impotence of public planning has meant that Houston's citizenry witnessed privatized planning long before it became chic in national policy circles in the 1980s. When long-range planning has been viewed as necessary, it has often been carried out under the auspices of the business elites, such as the "Suite 8F crowd," or by important committees of the Chamber of Commerce.

HOUSTON'S ECONOMIC DECLINE

Houston has been called the city that never knew the Great Depression of the 1930s, but the 1982-1988 oil-dollar recession proved that Houston was not immune to oscillations in the business cycle. Between March 1982 and March 1983 economic activity declined and unemployment rose significantly in Houston. By the mid-1980s the recession was generating more hardship in the city than in the nation at large. For example, the unemployment rate grew more rapidly than that for the U.S., exceeding 10% by September 1983. Nonfarm payroll employment peaked at nearly 1.6 million in March 1982 and then posted declines in 16 of the next 17 months; a 10% drop in employment. The city was again hit hard later in the decade when the unemployment rate rose to 10.3% during 1986 (9% for 1987), much higher than the national average. By January 1987, Houston had 221,900 fewer jobs than in March 1982. In addition to job losses, Houston also suffered large numbers of business failures.

The oil economy of Texas has been pivotal to the state's economic vitality. But the state's oil industry is changing. The industry confronts both short-term and long-run problems. The long-term problem is that oil reserves are progressively being depleted. Between 1972 and 1983, approximately 6 billion barrels of oil reserves were discovered, substantially less than the 10.6 billion barrels pumped during that period. Most Texas oil comes from huge fields discovered decades ago, and most recent discoveries have been small by historical standards. By the mid-1980s, Texas oil reserves were only half of what they were in 1952 (Jankowski, 1986, p. 27). If, as some have suggested, Texas' oil rigs are silent in 20 to 30 years, then the decline in the oil industry will severely affect the state's economy, workplaces, and educational institutions.

In the short term, the problem has been a sharp drop in the price of oil in the early and mid-1980s, followed by a period of unpredictable fluctuation. For a time lowered oil prices enabled consumers to buy fuel for less than one dollar a gallon. After plummeting below $10 a barrel in the mid-1980s, Intermediate West Texas crude oil had increased to nearly $21 a barrel in early July 1989 (see "OPEC," 1989, p. 1). Still, this price was down sharply from the $30-plus per barrel the crude cost in the early 1980s. Changes in the price of oil have disparate effects on the state and national economies. A drop in oil prices puts the brakes on inflation by reducing the raw material cost of many products and services. But this positive impact for Americans as consumers spells adverse consequences for Houstonians and Texans.

Another indicator of the Texas economic crisis in the 1980s has been a fragile financial sector. In addition to oil and gas companies going bankrupt, Texas has also endured many bank failures. In 1986, 26 banks in the state failed. By the following May another 23 had collapsed (Tolley, 1987, p. 1). Only the Great Depression witnessed so many bank failures. Of the 10 largest failures involving Houston financial institutions, six occurred between March 1986 and May 1987. Among the casualties was United Bank-Houston, the largest failure in the city's history (O'Grady & Seay, 1987, p. 1).

In 1982 there was a net in-migration of 417,000 people to Texas; by 1986, this had dropped to fewer than 5,000 people (Hobratschik, 1986, p. 9). Again, the crisis was more pronounced in Houston. The loss of jobs reduced net migration to Harris County to less than 7,000 in 1984 from nearly 49,000 in 1981. And 1985 began the first of four consecutive years in which out-migration would exceed in-migration. During that period, Harris County lost an estimated 125,000 people (Greater Houston Chamber, 1989a, p. 5).

STRATEGIES TO REVITALIZE HOUSTON

Houston's elites were slow to respond to economic decline and they were slower to recognize the full range of its causes. As recently as 1982, the State of Texas was spending 11 cents per capita in funding economic development efforts, compared to 94 cents for Louisiana, 63 cents for Oklahoma, and 68 cents for Arizona (See "Money," 1982, p. 4D). The reason for this lackadaisical approach is that prior to the 1980s, economic development seemed to take care of itself. The economies of Texas and Houston have generally expanded steadily since the turn of the century. Because of market-generated growth, local elites have concentrated on creating and perpetuating the "good business climate." However, confronted with a declining energy-based economy and a growing interconnectedness with international events, many business and political leaders have finally begun seeking more publicized governmental involvement, particularly in regard to diversification. Increased state-funded, economic-oriented activity—typically through the mechanism of "private-public partnerships"—has been seen as the way to diversify the Texas and Houston economic bases.

Houston's business leadership movement in the direction of an economic development program aimed at diversifying its oil-agricultural base was tardy compared to the 15,000 programs found in communities across the country in the early 1980s. The leaders of the Houston Chamber of Commerce established the Houston Economic Development Council (HEDC) only in 1984. This was the first explicit economic development organization in the city's history and the first devoted to aggressive corporate recruitment for economic diversification. With just two exceptions, the HEDC's chair and vice-chairs were drawn from its real estate-oriented growth coalition, which includes bankers, developers, and newspaper editors (See "City," 1984, pp. 9-13).

Broadly, the HEDC has sought to both generate diversity in Houston's economy and market the city more effectively. In 1984 the organization identified nine pivotal areas regarding economic development and diversification. But few of these had firm roots in the Houston economy. Included in the list were biomedical research and development (linking NASA and the medical center), research and development labs, and instruments. Notable omissions were oil, petrochemicals, construction, and most manufacturing industries. Ostensibly, the huge job losses in these areas during the 1982-1986 recession in Houston accounted for their omission. Many in the HEDC believed the new target areas had better prospects than traditional ones associated with the Texas economy (Clark, 1985, pp. 9-10).

The targeted areas were immediately criticized. Some businesspeople and media members pointed out that there was little precedent for establishing biomedical research as the top priority. Prior to 1986 Houston's predominantly treatment-oriented medical center had not spun off a major medical technological firm. Indeed, the area's largest med-tech firm, Intermedics, is based in the Angleton/Freeport area. While this company regards the Houston medical complex as a major customer, much of its developmental work was completed in collaboration with Massachusetts and California universities and electronics firms (Clark, 1984, pp. 37-38).

Furthermore, the Center for Enterprising at Southern Methodist University conducted a study suggesting Houston's leaders should direct more attention to diversification, that is, to basic product areas such as food, printing, pharmaceuticals, and cleaning products. The SMU study labeled biotechnology, space technology, and computers as less stable, faddish areas. A local citizens' group, the Greater Houston Tax Coalition, raised questions about HEDC goals and openly opposed the use of public funds for the organization. The coalition pressed for more education, reducing taxes and regulation as well as building on Houston's traditional strengths (Crown, 1986).

THE PRIVATE-PUBLIC STRUCTURE

The HEDC's first president was Don Moyer. Later, HEDC officials appointed a longtime Houston business leader as president of the organization. The executive selected, Lee Hogan, was a former owner of Universal Electric Construction Co. Inc., a local engineering and construction firm. Unlike Moyer, Hogan had no experience in economic development, but defended his appointment by arguing that the HEDC now is "at a different level in the stage of development." Though Hogan conceded Moyer's experience in economic development may have been necessary "to properly structure the organization," Hogan believed his appointment was in keeping with the HEDC's new agenda of "pulling the community together" (Sablatura, 1987, p. 1).

In the spring of 1989 the leadership at the HEDC changed again when John R. Brock was named president and chief operating officer. Brock had held the position of senior vice-president of marketing for Paine Webber Properties, Inc. Hogan, who became president of the Greater Houston Partnership (see below) said of his successor: "With his background in marketing, finance and real estate, he will provide important direction and input in our sales and marketing programs for Houston" (Greater Houston Partnership, 1989a, p. 1). In 1989 Brock stressed that the HEDC's primary activities were in three major areas: small business, sales, and marketing. He added that there were five target industries he believed would be productive for Houston: the Japanese marketplace, chemicals, aerospace, financial services, and California-based companies. He also lauded the support given to the HEDC by the Houston business community: "I believe that the tremendous support we received from the business community on various projects has been the major contributor to HEDC's success" (Greater Houston Partnership, 1989b, p. 8). Brock's comments were more revealing than he probably realized. The emphasis on the business community's involvement in Houston's private-public partnership demonstrates how little grass-roots groups have participated in the organization.

At the time of Brock's appointment, the Greater Houston Partnership was being created. As an industry publication indicated, "With the addition of the Houston World Trade Association, the Greater Houston Partnership has brought together the major organizations involved in Houston's business advocacy activities" (Greater Houston Partnership, 1989b, p. 2). The Houston World Trade Association claims it is dedicated to "providing services and networking for international business development." It is one of four groups shaded by the Greater Houston Partnership umbrella. The others are the Greater Houston Chamber of Commerce, devoted to programs such as regional systems; the Houston Economic Development Council, specializing in "target marketing, sales and business development"; and Partnership Shared Services, which provides research and other types of services for the rest of the Greater Houston Partnership (Greater Houston Partnership, 1989b, p. 2).

At the time of the restructuring, former HEDC president Lee Hogan said it "demonstrates the concept of one umbrella organization—the Greater Houston Partnership—overseeing Houston's business advocacy" (Greater Houston Partnership, 1989b, p 1). Hogan also stressed that the realigned structure provides "a focal point for setting public policy on business climate, business activity, job creation, economic development and quality of life issues" (Greater Houston Partnership, 1989b, p. 1). Hogan believes the revamped structure will be more effective in lobbying for state government assistance than the previous attempts by scattered groups with divergent agendas (Greater Houston Partnership, 1989b, p. 1). According to the partnership's executives, they speak for all of the business community. But what kind of mandate can the Partnership really claim? Our study suggests that opinion is not uniform among the various business factions in Houston, much less among the various non-business groups in the broader Houston community. All of Houston's business leaders have not embraced the activities of the Houston Partnership and the HEDC.

Although city growth coalitions typically draw up economic development plans and priorities before presenting them to the general public, the HEDC was created long before a broad vision was expressed. Indeed, in its first two years the organization spent most of its time fund-raising and trying to articulate a plan of action. Altogether, the HEDC raised $6.3 million, most of which was deductible as "business expenses." The organization spent most of its first two-year budget on consultants and other salaries. It also spent much for rent and improvements to its downtown office and for travel and entertainment (Gravois & Seay, 1986, p. 1).

Even though the HEDC is a privately conceived organization, city and other local public agencies have played an expanded financing role. Initially, the Houston City Council contributed $150,000 to the HEDC's first budget; the Port of Houston Authority pledged $75,000; and the county government contributed $500,000 (HEDC, 1986a, p. 4). Then, in March 1986, the council unanimously voted to match private contributions to the HEDC. Mayor Whitmire resisted, arguing it would mean higher taxes, and attempted to defeat the measure by attaching a minority business participation rider. But the HEDC proponents accepted the amended program. Further, the city controller later "found" $1.9 million in unused money, which was transferred to support the HEDC. The Greater Houston Political Action Committee, a business PAC, worked hard to get the grant passed. Members of the local business elite called it the "first true economic wedding of the public and private sectors" (Hart, 1986, p. 1A).

In many public-private partnerships, public sector officials have taken the lead and subsequently invited the business community's participation. But the HEDC has not adopted this approach. Rather, the HEDC is a private organization, funded with private money. Importantly, the public role has been restricted largely to financial support. In its 1987-1988 budget, about one-third of the funding came from contracts with local public bodies, including $1.25 million from the city of Houston, and $700,000 from the Harris County Industrial Development Council (O'Grady, 1987, p. 1F).

Some HEDC leaders have expressed fears that financial ties to government mean a closer public scrutiny of HEDC operations and expenditures (Hart, 1986, p. 1A). The HEDC officials, including then-president Moyer, worried openly that, if the HEDC accepted $1.25 million in grants, it would have to operate in a more public fashion. Moyer noted that corporate officials prefer to keep the decision-making process secret. He expressed the hope that elected officials and journalists would respect "the sacred character of our business transactions" (Gravois, 1986, p. 1A). Moreover, three years later, (two years after his departure) Moyer underscored the preoccupation with financial privacy when he said, "one of the nice things about being in the private sector is that when I left HEDC, one day later, my books were my own" (Miller, 1988, p. 11).

Many critics have questioned the HEDC's use of money. Lynn Ashby, a vice-chairperson of the HEDC and an editorial executive with the *Houston Post,* sums up the organization's image problem: "Houstonians are asking two questions: What's the HEDC done? Where's the money going? . . . the perception is growing that the HEDC is not a very tightly run business" (Ashby, 1986, p. B1). While Ashby remains sanguine about the HEDC, some neighborhood leaders, along with some former business supporters, have grown anxious about the organization and its activities. For example, some HEDC-associated business leaders have criticized excessive salaries and extravagant trips that have provided luxurious accommodations for participants, but no results. Among the issues that have caught insiders' attention is the salary of HEDC's president—$135,000—in 1986. City Council member

John Goodner has publicly questioned the wisdom of high salaries in a period of budget deficits and economic stagnation. For example, the organization's 20 full-time employees averaged $37,725 in 1986, more than double the salary of $18,000 for City of Houston employees. Council member Goodner and others also question the April, 1986, trip taken by a 17-member Houston delegation to Europe, which cost more than $250,000 (including $10,248 for limousines and $300 daily hotel charges). HEDC's officials admitted no tangible results were produced from the trip (Gravois & Seay, 1986, p. 1A).

Other business owners who originally supported the HEDC now question the continued funding of overseas trade missions and the "Houston Proud" campaign. This sentiment may be fairly widespread, but businesspeople are reluctant to voice their reservations publicly. One was quoted as saying "it's dangerous to criticize the HEDC" (Gravois & Seay, 1986, p. 1A). Thus, despite the emerging criticism, HEDC's president Moyer, Mayor Whitmire, and other HEDC officials took a two-week marketing trip to Japan in March 1987 to pursue firms in the Pacific Rim (See "Local," 1987, p. 3A).

Hoping to polish the city's (and perhaps the HEDC's) image, much of the HEDC budget has been spent for communications, public relations, and marketing efforts (HEDC, 1986a, pp. 13-15). The business elite has attacked both local and national media for portraying Houston negatively. An HEDC survey revealed that executives inexperienced with Houston viewed the city as a low-tech, single-industry "cowtown." The survey results were central in motivating the HEDC to redirect its priorities (HEDC, 1986c, pp. 8-9).

One attempt to reverse Houston's image was HEDC's retaining of New York and Chicago public relations firms to advertise the city. In 1985 the Chicago-based firm sent letters to 120 media outlets, protesting the city's negative portrayal and praising its virtues (Curtis & Novack, 1985, p. 43A). In addition, the HEDC coordinated the "Houston Marketing Network" to recruit local executives to extol Houston's advantages to their associates nationwide. In the organization's own terms, the network is "a volunteer 'sales force' of Houston business people who alert HEDC to companies considering business relocation or expansion." The network also "disseminates the HEDC's business message about Houston while selling their own product, service, or company" (HEDC, 1989b, p. 4). Another program, "Houston Beautiful," is a beautification campaign. Yet another, "Houston on the Move," is designed to heighten awareness among local residents about the $1 billion the city has budgeted to improve transit problems (Sheridan, 1986, p. 95).

And the "Houston Proud" media campaign was inaugurated. According to the HEDC (1989b, p. 4), Houston Proud "involves all interested citizens in economic development." Mainly a propagandistic device, the program was crafted to make Houstonians think more positively about the city. According to its organizers, Houston Proud is the most visible commercial campaign in the city's history. But not everyone is convinced of its value. Rice University sociologist Stephen Klineberg calls the $750,000 budgeted for the program a "waste of money" (Gravois, 1986, p. D2). Other business leaders are critical of the program, but again fail to speak out for fear of being portrayed as anti-development. One member of the local business community who has not been shy about speaking out is Edward Pita, president of Cabinet Systems, Inc. He calls the media blitz "a rather insulting and childish campaign," and adds "that the money should be used to develop businesses" (Gravois, 1986, p. D2).

THE REGENERATION POLICIES

In addition to establishing a unique semi-entrepreneurial private-public partnership, the business elite in Houston has continued to stress the "good business climate." An HEDC brochure, titled "The Houston Business Advantage," lays out the city's reasons for its reputation as a favorable business climate. Ironically, some of the advantages outlined stem from the vestiges of Houston's crisis. It mentions that:

> Texas is also a right-to-work state, with a labor force unionization rate of only 12.5 percent—the ninth lowest in the country. . . . Affordable Office Space. Approximately 25 to 35 percent of the city's land, office, warehouse and industrial space is available for immediate occupancy—at leasing and purchasing rates that are the most competitive of any metropolitan area; housing costs that are 20 percent below the national average; . . . Low Taxes. . . . Your business pays no corporate income taxes—city or state—no unitary tax and no state property tax. Tax abatements with grant reductions of up to 100 percent are available for up to seven years [HEDC, 1989a, pp. 2-3].

According to the HEDC, Houston's tax abatement policy attracted at least seven expansions or relocations to the city in the first six months of its latest incarnation. These seven companies, including Arco and Chevron, plan to create at least 1,700 jobs. Five of the companies that said abatement was one of the reasons for their decision to locate in Houston were chemical companies (HEDC, 1988, p. 2). That the majority of companies were in the chemical industry raises the question of just how critical the abatements were to the investment decision. The recently revamped policy specifies that companies that create at least 30 jobs and make capital improvements of at least $5 million are eligible for an abatement. Chevron was an early recipient of the new Harris County policy. In February 1988 the company announced it would build a $100-million polyethylene plant at its Cedar Bayou facility in Baytown, creating an estimated 70 permanent jobs and 750 temporary construction jobs. In all, the abatements Chevron received from Harris County, the City of Baytown, the Baytown Independent School District, and the Lee College Board of Regents were worth $1.7 million annually over a five-year period.

Another part of the HEDC's stated agenda is to improve support for new businesses. The 1986 Strategic Priorities Agenda recognizes that the city has no state venture capital funds, that many business sectors have been redlined by local and national banks, and that "public sector sources of business financing are very limited" (HEDC, 1986c, p. 5). The HEDC solution relies on "incubators without walls" and public sector financing backed by the Chamber. "Incubators without walls" refers to the HEDC's Direct Business Assistance (DBA) program, which is designed to give entrepreneurs access to professional services through a referral system. The HEDC spent $100,000 to develop the program and hired a director to initiate it. On the surface, such contributions appeared to pay high dividends. According to *Site Selection* magazine, the HEDC is among the top economic development groups in the country (Cited in Greater Houston Partnership, 1989b, p. 1). The magazine cited the 26 businesses that located or expanded in 1988 in the Houston area, adding 6,500 jobs and precipitating an economic impact of $1 billion, as evidence of the group's entrepreneurial skills. One of the reasons the magazine designated the HEDC as part of its "cream of the crop" was its ability to persuade Compaq Computer to stay and expand in Houston. This Compaq success story was lubricated by a multimillion-dollar package of incentives, including training programs, recruiting assistance, tax breaks, and roadway expansions.

The variety of subsidies for business has continued to grow. For example, Houston Light and Power offers an economic development incentive that discounts electric bills of qualified users by an average of 20%. In effect, the utility freezes the firm's electric bill for five years, saving the typical light industrial or commercial user about $150,000 over a five-year period. Chemical companies benefit disproportionately; the typical chemical firm will save about $4 million during five years (HEDC, 1988, p. 2). According to the Edison Electric Institute and the National Economic Research Associates, Houston now offers large electricity users one of the lowest rates in the country (HEDC, 1988, p. 2). To qualify, a company must "demonstrate to HL&P that their facility is a basic addition to the Houston economy" (HEDC, 1988, p. 2).

Beyond these industrial investment strategies, the HEDC attempted to secure a Navy homeport for the Houston-Galveston area. The HEDC's private sector orientation has not precluded its pursuit of state-funded projects, such as military bases. Galveston officials and the HEDC packaged $18 million in incentives to try to lure a proposed naval base (home port) for a U.S. battleship group. The naval base (eventually awarded to Corpus Christi) would have meant $100 million in construction contracts, 2,400 construction jobs, and a $60-million annual payroll.

It may well be that Houston's "good business climate" will do little more than guarantee the impoverishment of the local state because of its heavy reliance on expensive tax giveaways to lure new employers. Clearly, this is not the only option available to Houston. One major alternative is to tax more and provide a higher level of government services. Despite the number of financing mechanisms that Houston has available to it, elected officials there historically have found revenue enhancement politically inexpedient. Houston's capacity to exercise alternative options would have been easier if the local state had traditionally represented a more diverse constituency. But in this city, there is no active political contest driving wedges between economic elites. There is no strong, organized opposition to present a comprehensive challenge to the status quo. Subsequently, Houston leaders operate in a reactive mood, trying to stave off the worst effects of economic downturns. Much of the HEDC's focus has been on narrowly drawn business concerns. But the HEDC and the Chamber have been forced to acknowledge that, at least in mass transit and education, the public sector may need attention.

Traffic problems have long been symbolic of the underside of Houston. Congestion problems were so serious by 1978 that many business leaders came out in favor of government-subsidized rail mass transit. A study by the American Public Transit Association reported that Houston had one of the worst public systems in the United States. In a local election Houstonians approved the creation of a Metropolitan Transit Authority (MTA), to be funded by a 1% sales tax, with a mandate to plan a rail transit system. In 1982 Houston's MTA approved the construction of a $2.1-billion heavy-rail system to run from downtown out the southwest corridor, to help solve severe traffic problems in that booming area. Business leaders were quite willing to accept federal funding for the transit plan. As is usually the case, this early-1980s plan was designed to meet business needs. The proposed rail system would guarantee that much of the city's future development would be centered in the downtown area. And the system would connect the downtown center to the key outlying business activity centers, such as the

Post Oak and Greenway Plaza areas. This heavy-rail proposal was vetoed by the voters, and the city authority switched to a light-rail concept in the late 1980s.

Increasingly concerned with the city's image, Houston's leaders have milked their improvements in mass transit for public relations purposes. According to a report that was partly sponsored by the Chamber, traffic congestion in Houston dropped 5% between 1984 and 1986, while most other large urban areas experienced increases. An executive with the Chamber's Regional Systems division claimed that Houston has made the most significant progress in its transportation programs since the study was completed. He said the biggest factor has been the massive road building program since 1984 (Greater Houston Partnership, 1989a, p. 2).

However, it seems just as likely that any reduction in traffic congestion is attributable to a decline in economic activity and a net loss of population. But these factors are conveniently overlooked. In a recent Chamber publication, a "Traffic Congestion" chart showed Houston's situation in a favorable light compared to Los Angeles and Washington, D. C. As the head of the MTA board said in the early 1980s, "nobody paid attention to transit here for 20 years or more. Our bus service was probably the worst in the country. We got so much growth, people did not prepare; people were complacent" (Reinhold, 1982, p. 11). By the late 1980s, however, this administrator had sharply reduced system accidents, bus failures, and costs. Meanwhile, the number of passengers increased significantly, to 74 million in 1987.

The national media have amplified the news about the improvement in Houston's transportation system. According to *Time,* "Houston now boasts a highly efficient transit system that the American Public Transit Association ranks as the safest in the U.S. The buses are on schedule 98 percent of the time and are so dependable that they need repair only once every 11,000 miles, compared with the U.S. average of 4,000 miles (See "Fast," 1988: GHP reprint). Still, one should not exaggerate the progress, for only 3% of all Houston trips now take place on a city bus.

In the late 1980s, with business elite backing, funding for a substantially revised mass transit plan, one that met not only downtown needs but also those of a larger segment of the Houston community, was finally approved by Houston voters. The HEDC and other business organizations eventually learned that elite goals had to acknowledge, even if minimally, the demands of the general public.

Education is another area where Houston's business elite has recently paid attention to major increases for broader social programs. The concern with education was signaled by an article in the Chamber's *Houston* magazine, titled "What can Houston Learn from Massachusetts?" The article belied

Houstonian's traditional disdain for Yankees and their ideas. It praised a symposium, which featured Massachusetts officials and educators lecturing to Houston business people. The basic theme was that the presence of major research universities in Boston had created the large reserve of innovative researchers and potential entrepreneurs necessary for economic vitality. Massachusetts officials revealed that federal- and private-sponsored research at Rice University and the University of Houston reached $34 million in 1985, compared with $250 million for MIT. Like many local critics, the Massachusetts officials argued that Houston's leaders should avoid high-tech electronics and build on traditional areas, on NASA, on chemical and mechanical engineering firms, and on medical institutions (Clark, 1986, pp. 9-13).

The attempts to resolve the economic crisis, the multiple sources of criticism, the symposium, and a confluence of related developments left its impact on the HEDC. By mid-1986 officials were speaking about the "essential" role of higher education in economic development. One HEDC executive member said, "We feel that enhancing higher education represents the greatest opportunity to strengthen and diversity our economy," and that developing the University of Houston is "essential for our progress as a world-class city and the economic development we all aspire to" (Loddeke, 1986, p. B5).

Important differences remain between the HEDC and its parent—the Greater Houston Chamber of Commerce—about how these objectives should be met. For example, HEDC executive member John E. Walsh, Jr. (president of Friendswood Development Company) believes that increased taxes for higher education may be required to "achieve a stable, more diversified economy." The Chamber, whose subsidiaries accepted public funding, has instead stressed corporate investment in education. Jonathan Day, a Houston law firm partner and chair of the Chamber's educational task force, says his organization will act as a catalyst to encourage area businesses to donate more money to schools. He called private sector support "critical" because of diminished state resources (Loddeke, 1986, p. B5).

For its part, the HEDC proposed the creation of a community-wide task force to strengthen research and graduate training in local universities. Regardless of the faction to which they owe their allegiance, Houston's business elites are careful about how they promote education. Generally, they favor the short-term investment strategy approach. For example, while the state's university system remains first-rate, the public schools are among the most deficient in the nation. If faithful to their rhetoric, this would be the area where the financial commitment of the local elites would be most evident. Supported by the business elite, a recent billion-dollar bond issue was passed for the Houston Independent School District. While modest in terms of the tremendous backlog of educational needs, this is a first step toward revitalizing primary and secondary education in Houston.

The economic development strategy selected by the city's HEDC and its political allies has had a deleterious impact on the public infrastructure in Houston. The examples of increased attention to mass transit and education just cited are the exceptions rather than the rule. Nonetheless, if more members of the business elite decide to move in a modestly progressive direction, it may yield a new definition of the good business climate. For Houston to maintain its incremental climb upward, it must reinvigorate social service programs like mass transit and education, programs that will mean tax increases. Then the area will be seen as desirable to organizations seeking better government services and a higher quality of life, like the ones usually listed in the annual business climate rankings of Grant Thornton and others. By all traditional accounts, Houston has a very good business climate, but the definition of a good business climate varies; Houston's low-tax, poor public service represents an extreme version.

The impact of the impoverishment of local services is obvious. Consider lax air pollution standards. Poor public services means Houstonians are regularly exposed to hundreds of pollutant-spewing industrial plants and motor-vehicle smog. Traffic congestion in Houston (a mid-1980s Federal Highway Administration study of 37 metropolitan areas ranked Houston last in traffic) has also contributed to a high death rate. In 1980, Houston's traffic fatality rate of 23 per 100,000 per year was twice that of Detroit and nearly twice that of New York and Philadelphia. The sorry state of public infrastructure does not end with traffic problems. In the mid-1980s, Environmental Protection Agency head William Ruckelshaus described the metropolitan Houston area as one of the worst hazardous waste disposal areas in the nation because of the nine major toxic waste dumps then on the Superfund list. He said the area has "bigger problems because it has more of the industries [that produce waste]" (Feagin, 1988, p. 214).

Houston's low-tax/poor public service approach has also produced a series of interrelated water problems. For example, the strategy has rendered sewage and garbage disposal problematic. The garbage disposal problem has escalated the number of intracity conflicts over where to locate new disposal sites. By the mid-1980s millions of gallons of wastewater were flowing into the long-since polluted Buffalo Bayou from city and private-development sewage plants. In some places, the Bayou has become anaerobic, with little aquatic life remaining.

Other aspects of Houston's water problems are sufficiently unique and so severe that some parts of the city are sinking. The Houston area's dependence on drawing the majority of its water supply from wells and from lake reservoirs has created changes in soil composition leading to a problem known as subsidence. Subsidence has caused places like southeast Houston to sink six to eight ft. since 1943, with the central city having subsided three to five ft. in the same period. The continued existence of adequate water supplies is another question mark. Supplies are supposed to be adequate until 1992, but there are already occasional problems with water pressure. Periodic flooding is yet another type of water problem besetting Houston. Average annual losses due to flooding in the 1970s and 1980s were running more than $30 million.

No water problem is as grave in undermining the aesthetic appeal of Houston as the deterioration of surface water in the city. Perhaps the most prominent case is Lake Houston. After finding that Lake Houston's fecal coliform count was more than 15 times the "safe" standard, an environmental specialist cautioned city residents to avoid the lake. Of the many factors contributing to Lake Houston's pollution, real-estate investment and development is critical. In the mid-1980s between 140 and 160 million gallons of water from this lake were being treated daily at a treatment facility with a rated capacity of only 100 million gallons. Large corporations have also played a major role in Houston's water pollution problems. In one instance, the E.P.A. ordered privately owned Houston Lighting and Power to cease discharging millions of gallons of untreated wastewater into lakes and streams near its coal-fired plants.

THE HEDC AND REVITALIZATION

Despite the efforts of local business elites, Houston was still reeling in the late 1980s. One area that showed little sign of expansion was the energy infrastructure industry. Having hit a postwar low of 686 active rotary rigs in July 1986, the number of rigs operating in April 1989 was a meager 763 (Greater Houston Partnership, 1989a, p. 6). The construction industry has also resisted recovery. In addition to these trouble spots, Houston and Texas were still beset with hundreds of insolvent financial institutions. And though the rates were increasing, the real estate industry continued to be burdened by low occupancy levels, particularly in big office projects in suburban areas.

By early 1989 Houston's business leaders were boasting about a number of economic indicators: a 4% percent employment growth rate; the creation of 44,000 new jobs in 1988; a manufacturing employment growth rate exceeding 4%; an unemployment rate of 6.6%; an inflation rate of 3.7%; a foreclosure level 37% below the previous year, a business failure rate 14.5% lower in 1988 compared with 1987, and a 12% increase in Port of Houston tonnage (HEDC, 1989d, p. 1).

The HEDC officials have claimed credit for enticing new companies to Houston and for the expansion of existing facilities. In all, the HEDC stated they helped secure 26 new companies or new company expansions in 1988. The HEDC's biggest victory was in convincing Compaq Computer to stay and expand its facilities. HEDC's officials formed a task force of 75 Houston leaders and went to Compaq executives with one question: "What will it take to keep you here?" (Rose, 1988: GHP reprint). The incentive package was expected to total $235 million and included, among other items, a bus service plan; $15 million in tax abatements, $15.5 million in electricity discounts; publicly funded customized training programs for new employees; and a promise to widen a local road into a 10-lane, limited-access highway. At the time, the company's president was quoted as saying: "They blew us away. I've never seen a city move that fast" (Ivey, 1989, p. 103). Lee Hogan said of the agreement, "If I had to write a textbook on how to go about economic development, that would be case history No. 1" (Rose, 1988).

Beyond the activity generated by the HEDC's foray into entrepreneurial development, Houston's economy was showing other signs of revitalization in the late 1980s. One area of regeneration has been the petrochemical industry. Houston's petrochemical revival is part of a national trend, independent of local actions. Petrochemical exports from Texas in 1984 totaled $3.9 billion, representing 35% of the value of all manufacturing. In the first quarter of 1988, the industry employed 45,749 workers (Koenig, 1988). And the future looks promising. According to a recent survey by the HEDC and Houston Light and Power, nearly 150 petrochemical expansion or construction projects worth $5.5 billion were planned for Houston. Ray Porter, an economic development analyst for HL& P, says that by "1992 the Houston Gulf Coast area could see 174 projects worth almost $7 billion." The HL&P executive added that U.S. firms are responsible for 70% to 80% of the petrochemical projects (*Journal of Commerce*, 1988b).

Trumpeting the "good business climate" of Houston, Saadat Syal, an executive at the HEDC, said many of the planned expansions were attributable to Houston's existing chemical industry infrastructure, the availability of a trained labor force, and the city's transportation network (HEDC, 1989c, p. 2). But the petrochemical industry demonstrates the importance of international factors shaping Houston's economy. Thomas Plaut, senior economist of the comptroller's Economic Analysis Center, notes that "the dollar's 30 percent to 40 percent drop from the middle of 1985 to the beginning of 1987 helped revive the export business by making U.S. petrochemical products more competitive in world markets" (*Journal of Commerce*, 1988a). In short, a large part of the petrochemical industry's current vitality is due to factors beyond the scope of the HEDC.

Though minor, the HEDC has played an economic expansion role in the aerospace industry. In 1962 the National Aeronautics and Space Administration selected Houston as its center for manned spaceflight. The decision (a product of the close links between local business elites and national political leaders) established the city as an integral component in the nation's space exploration program. It also set the stage for private companies seeking to capitalize on the commercial application of space technology. The Houston NASA center has primary responsibility for the research, design, development, and testing of large parts of the space shuttle and the space station programs. Well-established Houston companies, such as Brown & Root and Cameron Iron Works, have long had space-related contracts with NASA. According to Rice University sociologist Stephen Klineberg, Houston has prospered on the base of two economic pillars, oil and space: "We knew oil would run out some day, but we felt secure that the space industry would carry us through . . . oil was the industry of Houston's past, and space was the industry of Houston's future" (Belkin, 1988).

With the resurgence of spending in the space program, a host of contractors are showing signs of interest in the Johnson Space Center. One indication of this reinvigorated sector is the decision by Westinghouse to invest in Houston-based Space Industries Inc. (Dennis, 1987, p. 46). But Grumman Corporation's plans not only are more widely touted by the city's business elites, but also attract more media attention. In 1988 Grumman was in the early stages of constructing a manufacturing center for the Space Center. If events unfold as planned, it will employ 2,000 workers when it opens in 1992. Grumman's expansion was greased by an incentive package that included free land for 20 years and tax abatements worth $7 million (O'Connor, 1988). There are more than 80 companies in Houston involved in space-related activities.

With the exception of Grumman's expansion, little credit can be given to the Greater Houston Partnership, or to the HEDC specifically, in stimulating economic regeneration in space-related industries. The actions of Houston's business elites *were* critical in having the city initially assigned as the center of manned spaceflight projects, but the recent resurgence is more attributable to decisions made at the national state level than at the local state or business elite level.

ENTREPRENEURIALISM AS BEGGING

Houston's business elite was one of the first to invent the entrepreneurial approach to urban development in the nineteenth century and has persisted in that vein ever since. Indeed, one of the 1836 advertisements for the city by the Allen brothers said the following:

> The town of Houston is located at the point on the river which must ever command the trade of the largest and richest portion of Texas. . . . [it] will warrant the employment of at least *One Million Dollars* of capital, and when the rich lands of this country shall be settled, a trade will flow to it, making it, beyond all doubt, the great interior commercial emporium of Texas [Feagin, 1988, p. 111].

These entrepreneurs solicited outside capital and offered the land to investors, merchants, and settlers, with the intention that Houston would become the regional center of commercial capitalism and the government of the new Texas Republic.

Recent HEDC development efforts are in some ways the modern-day incarnation of this entrepreneurial pursuit of outside capital. In particular, they reflect the first considerable concern over the diversification of the local economy in many decades. From 1900 to the early 1980s the oil-rooted economy was firmly established under the guidance of a succession of business elites. It was this oil monoculture, however, that created problems for the city, as the price of oil declined and the value of the dollar rose in the 1980s. The HEDC was belatedly created to cope with this crisis of diversification.

However, HEDC activity has been relatively unsuccessful in redirecting the oil monoculture. HEDC's actions influencing the location decisions of a few corporations—some of which have been oil-related—generate much publicity but have several basic problems. First, they tend to exaggerate the influence of the HEDC, as opposed to the fundamentals of the local economy. Second, they are costly in terms of governmental expenditures, including support for the HEDC, and in terms of tax losses. City governments like Houston often "give away the store" for corporate location decisions that might have come their way in any event and whose benefits cannot be guaranteed over the long term. Third, these HEDC policies have, so far, had little impact in terms of increasing the total number of jobs in the city's economy: The number of direct new jobs created by the major relocations announced in 1988 totaled less than 0.3% of the *Texas* employment base (Newport, 1989). As Bratman (1988) noted, the 1988 moves created a modest 6,334 jobs. Even the chairperson of Houston's Commerce Department recently observed: "You can't build your future just by begging companies to relocate to Texas" (Newport, 1989).

The Houston economy was shaped by the entrepreneurial efforts of a succession of business elites, from the Allens to the more recent 1930s-1970s

"Suite 8F crowd" and finally to the Chamber of Commerce. Houston's development strategies are typical of many American Sunbelt cities. But the low-tax, poor-public-service development model fails to attract adherents in places like Pittsburgh, where political and economic leadership are more assertive in defining the city's future.

REFERENCES

Ashby, L. (1986, August 26). HEDC has image problem of its own. *Houston Post*, p. B-1.

Belkin, L. (1988, September 28). Houston counts on success by space shuttle to lift its sluggish economy. *New York Times* (GHP Reprint).

Bratman, F. (1988, December 16). Looking ahead: Houston will lead Texas real estate rebound. *McGraw Hill News* (GHP reprint).

City leadership presents economic development plan. (1984, July). *Houston, 55*, pp. 9-13.

Clark, R. (1984, May). Part two: High tech on the horizon. *Houston, 55*, pp. 33-38.

Clark, R. (1985, February). Development council maps economic strategy. *Houston, 56*, pp. 9-13.

Clark, R. (1986, May). What can Houston learn from Massachusetts? *Houston 57*, pp. 9-13.

Crown, J. (1986, April 2). HEDC sets goals for growth by year 2000. *Houston Chronicle*, p. 1.

Curtis, T., & Novack, J. (1985, March 24). Bad reputation. *Dallas Times Herald*, pp. 43-44.

Dennis, D. L. (1987, August 17). Trying to fix busted cities. *Fortune*, p. 46.

Fast lane to work. (1988, September 12). *Time* (GHP Reprint).

Feagin, J. R. (1988). *Free enterprise city: Houston in political and economic perspective*. New Brunswick and London: Rutgers University Press.

Feagin, J. R., & Parker, R. E. (1990). *Building American Cities*. Englewood Cliffs, NJ: Prentice Hall.

Gravois, J. (1986a, April 14). HEDC weighs strings attached to tax dollars. *Houston Post*, p. 1A.

Gravois, J. (1986b, September 21). Some critical of campaign but group still Houston Proud. *Houston Post*, p. D-2.

Gravois, J., & Seay, G. (1986, August 24). Spending practices of HEDC criticized. *Houston Post*, p. 1A.

Greater Houston Chamber of Commerce. (1989a, January). *Houston Economic Overview* (Prosperity 1974-1981), 3.

Greater Houston Chamber of Commerce. (1989b, January). *Houston Economic Overview* (Population), 5.

Greater Houston Partnership. (1989a, May). Brock named new HEDC president. *At Work*, 1.

Greater Houston Partnership. (1989a, May). Building a better city: Study shows road improvements ease Houston's traffic congestion. *At Work*, 2.

Greater Houston Partnership. (1989a, May). Economic indicators. *At Work*, 6.

Greater Houston Partnership. (1989b, June). HEDC named one of the nation's best. *At Work*, 6.

Greater Houston Partnership. (1989b, June). Partnership expands to include Houston World Trade Association. *At Work*, 2.

Greater Houston Partnership. (1989b, June). Partnership organization complete. *At Work*, 1.

Greater Houston Partnership. (1989b, June). An interview with the new HEDC president reveals insight, plans. *At Work*, 8.

Hart, J. (1986, March 24). Council wins fight to fund HEDC. *Houston Business Journal*, p. 1A.

Hobratschik, M. (1986, June 26). Officials expect Texas bank failures. *Daily Texan*, pp. 1-4F.

Houston Economic Development Council. (1986a). *Fund campaign report* (Typewritten). Houston.

Houston Economic Development Council. (1986b, January). *An overview of the Houston economy: Houston works for business report*, 1-3. Houston.

Houston Economic Development Council. (1986c). *Strategic priorities agenda* (Typewritten). Houston.

Houston Economic Development Council. (1988, July). *Houston business report: Corporate incentives*, 2. Houston.

Houston Economic Development Council. (1989a). *The Houston business advantage* (Pamphlet). Houston.

Houston Economic Development Council. (1989b). *Houston* (Pamphlet), 4. Houston.

Houston Economic Development Council. (1989c). Houston targeted for chemical expansions. *Houston Business Update*, 2. Houston.

Houston Economic Development Council. (1989d, March). *Houston: Back on top to stay*, 1. Houston.

Houston to see huge expansion in petrochemical industry by '92. (1988b, December 9). *Journal of Commerce* (GHP reprint).

Ivey, M. (1989, January 16). Houston's sick economy is taking a little nourishment. *Business Week*, pp. 102-103.

Jankowski, P. (1986, April). What's the status of Texas reserves? *Houston, 57*, pp. 27-29.

Jones, R. (1982). *Town and country chaos*. London: Adam Smith Institute.

Koenig, R. (1988, December 27). Chemical firms see slower growth in '89. *Wall Street Journal* (GHP Reprint).

Local task force heads for Japan. (1987, February 23). *Houston Business Journal*, p. 3A.

Loddeke, L. (December 10). College investing backed: Chamber seeking corporate support. *Houston Post*, p. B5.

Miller, D. (1988, September 19). Donald Moyer: Spirit of HEDC past. *Houston Business Journal*, pp.1-11.

Money matters: State lags in development funding. (1982, November 15). *Houston Post*, p. 4D.

Newport, J. P. (1989, March 13). Texas faces up to a tougher future. *Fortune* (GHP Reprint).

Nichols, B. (1988, July 27). Industry shakeout sends jobs to Houston. *Dallas Morning News* (GHP reprint).

O'Connor, M. (1988, October). Compaq, Grumman expansion plans spur Houston's rebounding economy. *Site Selection* (GHP Reprint).

O'Grady, E. (1987, February 28). HEDC surpasses goal, raises $7.2 million. *Houston Post*, p. 1F.

O'Grady, E., & Seay, G. (1987, May 1). Bank failure largest in city's history. *Houston Post*, p.1A.

OPEC production at year's highest level. (1989, July 6). *Las Vegas Review-Journal*, p. 1D.

Petrochemicals: A bright star in the Lone Star economy. (1988a, November 15). *Journal of Commerce* (GHP reprint).

Reinhold, R. (1982, September 28). Houston ponders public transit plan. *New York Times*, p. Y11.

Rose, C. (1988, December 26). Jobs return, optimism grows as Houston is weaned from oil. *New Orleans Times-Picayune* (GHP Reprint).

Sablatura, B. (1987, April 27). Who is Hogan and why did HEDC hire him? *Houston Business Journal*, p. 1.

Sheridan, M. (1986, April 1). Houston: A plan for diversification. *Spirit*, pp. 81-99.

Tolley, L. (1987, May 1). Bank failures bring Texas total to 23. *Austin American Statesman*, p. 1D.

Part IV

Searching for Leadership

12

Leadership and Regeneration in Liverpool: Confusion, Confrontation, or Coalition?

MICHAEL PARKINSON

REGIME INSTABILITY IN LIVERPOOL

Liverpool provides a classic location to examine the role of leadership in urban decline and regeneration. During the past two decades it has experienced a profound transformation under the impact of international economic restructuring, which has set before it major social and political challenges. However, a crucial feature of the city during this period is the way in which its leaders reacted to the challenges it faced. In many respects, the city's economic failure has been matched by a political failure that has exacerbated the costs of change. Ironically, the importance of leadership in urban transformation is illustrated by its absence—or incoherence—in Liverpool.

This book examines the capacity of leaders to create stable and durable mechanisms and alliances that promote economic regeneration and identifies a range of micro-level skills and macro-level resources that can generate that capacity. Of all the cities in this book, it is Liverpool's leaders who probably have the greatest deficit in both respects. They have failed to demonstrate the necessary political skills to construct a stable coalition to promote regeneration. But they also lack many of the resources that underpin leadership capacity. The public and private sectors are both weak, and political relations between them have been strained. Liverpool's relationship with central government has been controversial, and the city has attracted little national goodwill or money. The city's social structure is unpromising. It is dominated by a large working class and its culture; the middle class is relatively small. Class relations, mediated through political party and trade union action, have been tense. The educational and skill level of the community is low. There has been consistent emigration and little immigration to produce dynamic new social groups. The city lacks an entrepreneurial tradition. It does have some environmental potential but few locational advantages, especially since it lies in the shadow of Manchester, the more powerful regional capital.

The most obvious aspect of the city during the past two decades has been regime instability. For a variety of reasons connected to its peculiar economic and social development, the city has been characterized by highly volatile and partisan party politics, limited administrative and governmental capacity, a lack of powerful business leadership, and an inability to construct coalitions between the public and private sectors. In recent years Liverpool has lacked the stability and continuity among economic, political, and social leaders that have permitted some other provincial cities discussed in this book to respond proactively to economic decline and build the elements of a regeneration strategy.

For much of the period Liverpool was characterized by what Ann Markusen (1989) has termed "the politics of declining regions." The dominant feature of that pattern is that cities turn to external government agencies in a search for increased resources, rather than attempting to resolve economic problems internally. Liverpool had a limited degree of success with that strategy during the 1960s and 1970s, when public expenditure was rising and national governments of both political parties were expanding central funds to the cities. However, the strategy self-destructed in the 1980s, when it was pursued aggressively by a left-wing city council in the face of a powerful right-wing Conservative government determined to reduce central support for cities (Parkinson, 1990).

The recent history of Liverpool's political economy can be divided into three broad periods: 1973-1983, 1983-1987, and 1987-1989. They correspond to three major phases of political life in the city, which produced three different local economic strategies. The period from 1973 to 1983 witnessed a dramatic escalation of the city's economic problems, combined with a period of political paralysis because none of the city's three political parties could achieve the necessary electoral support to get a majority on the council and develop a coherent response to economic decline. The period from 1983 to 1987 was marked by the rise of a powerful Labour majority in the city, which regarded a major public spending program on the physical infrastructure in its working-class heartland as the only way to regenerate Liverpool's economy. Labour's tactics during this period alienated the Conservative government and the local private sector and ended in political and legal defeat for Labour (Parkinson, 1985).

After 1987, as the failure of Labour's strategy became apparent to all political actors in the city, an alternative development strategy began to emerge in Liverpool. During this crucial period, changes of leadership and strategies in the public and private sectors meant that the city's politics began to change from municipal socialism to urban entrepreneurialism. A Labour regime emerged that was eager to form alliances with the local private sector and pursue an economic development strategy more in tune with the priorities of national government (Harding, 1990). Liverpool began to make the strategic and political choices that the other left-wing cities discussed in this book—Glasgow and Sheffield—had made earlier in the decade.

LEADERSHIP SKILLS AND RESOURCES IN LIVERPOOL

When Liverpool entered its major phase of economic decline in the 1970s, it had a distinctive leadership structure that placed real constraints upon the city's capacity to respond. As with much else in the city, leadership patterns were intimately connected to its traditional economic structure and, in particular, the role of the port. The port dominated both the city's and the region's economy from the last part of the nineteenth century to the end of the Second World War, shaping its employers, its work force, its trade union structure, and its party politics. It gave it an employer structure that was dominated by large firms that, in the main, employed semi- and unskilled workers on a casual basis (Cornfoot, 1982; Lane, 1987).

This also meant that the local economy did not historically build up a base of skilled manufacturing workers who could transfer their skills to other sectors of the economy when the port began its decline in the post-1945 period. Those employed in the economy outside the port tended to be in the distributive and food processing trades. The dominance of food processing made the city especially vulnerable to the rationalization and centralization that occurred in the postwar period and led many local firms to be taken over by national and multinational corporations. During the 1960s the growth of a car industry in the region, which created 25,000 jobs in manufacturing, helped ease the city's immediate economic problems. But it reinforced the dominance of large multinational employers in the city's economy, which later turned to its disadvantage (Lloyd, 1979; Lloyd & Dickens, 1978).

The dominance by a limited number of externally controlled large firms not only made Liverpool's economy vulnerable, but also affected the structure of the city's economic elites. It meant that many of its largest economic actors primarily regarded themselves not as local firms with a stake in the local economy but as national and international actors with national and international markets. As a result they did not play a major part in the political life of the city. Nor did there emerge powerful coalitions of business groups with their own economic agenda for the city. Liverpool's major economic leaders were, at best, fragmented and failed to provide civic leadership.

The growing dominance of large firms during the postwar period also meant that Liverpool, even in comparison with other provincial cities, had less of a tradition of small, entrepreneurial firms producing local capitalists and capital. This also had an impact on the politics of the city since it limited the pool of indigenous capitalists who, in other provincial cities, frequently provide the political or civic leadership. Dramatic postwar demographic change including a massive hemorrhaging of more than 400,000 people from the city, allied to the suburbanization of many middle-class groups beyond the city boundaries, and also diminished the pool of economic leaders who might become potential political leaders in the city.

The economic and social structure of the city also affected the Labour party and trade union movement. The party, like the city, was traditionally dominated by working-class rather than middle-class interests and by blue-collar general trade unions rather than skilled-craft trade unions. This lack of middle-class groups and the dominance of working-class interests meant the party developed a leadership style and politics that were almost exclusively oriented to the needs of its working-class supporters and members. The class-based nature of the party also affected its internal political culture, which became dogmatic and workerist. Partly because of this, the party became vulnerable to alternative periods of political upheaval and militant insurgency, combined with longer, more stable periods of boss-led machine politics typical of many nineteenth-century American cities. In turn, those qualities emphasized the gap between the city's political leaders and economic leaders and increased the difficulty of forming a political alliance between the two (Baxter, 1972; Crick, 1986; Hindess, 1971; Mulhearn, 1987).

The port also affected the city's politics, since in the mid-nineteenth century it had attracted tens of thousands of Irish immigrants to the city. Catholic immigration divided the Liverpool working-class community by encouraging Protestant workers to support the Conservative party rather than Labour, which eventually became the party of the Catholic Irish working class. In this way sectarian, as opposed to class, politics were introduced into the city (Waller, 1968). It explained why the most proletarian city in England traditionally elected Conservative, not Labour, regimes. That pattern was not broken decisively until the 1980s, several decades after the other British cities in this volume were taken over by Labour.

The changing electoral fortunes of the Conservative and Labour parties during the 1950s and 1960s meant that, unlike that in many other British provincial cities, the city council did not experience a period of sustained political stability under Labour. In the 1970s this pattern reached its zenith, with a decade of minority and coalition governments in the city that fundamentally restricted its capacity to confront the fiscal, administrative, and social consequences of the city's rapid economic decline. The constant rise and fall of regimes meant Liverpool failed to take a series of difficult political decisions rationalizing the management and financing of the city. Politicians were unable, or unwilling, to confront the power of the blue-collar trade unions, which represented many council employees, and did not begin the rationalization of the city's basic services, especially in housing and environmental maintenance. The council was unable to modernize its administrative machinery, which became outdated and inefficient (Parkinson, 1985).

The politicization of the city council, which the instability produced, also discouraged a clear division of responsibilities between professional administrators and elected politicians, who often attempted to retain control of policy. This strained working relations between the two groups, making it difficult for the city to achieve clear administrative leadership. It also made it difficult to recruit able administrators from outside Liverpool, thus helping to prevent the emergence of a more dynamic administrative class in the city. The consequence was that as Liverpool entered the period of intense decline and needed to develop new economic strategies, the city council, which was the largest employer and the most importance economic actor in the city, was not capable of providing a clear political or administrative lead (Audit Commission, 1988).

1973-1983. LIVERPOOL'S LOST DECADE: ECONOMIC DECLINE AND POLITICAL FAILURE.

Liverpool's economic decline is intimately connected to the role of the port. At the beginning of the First World War it was the second-largest port in the British empire and one of Britain's richest cities. But changing patterns of technology, transportation, and communication from the 1920s onward sent the port into a long-term secular decline, which had a profound impact on the economy, in particular causing long-term structural unemployment. The port also had a major impact on the structure of the local economy, creating substantial dependence on semiskilled jobs and a blue-collar service sector in the city. The relative absence of a manufacturing base and of a white-collar service sector meant that the city was poorly positioned to expand into those sectors to compensate for the decline of port-related activity.

This long-term decline of the economy was dramatically escalated after 1973 by two economic forces—the oil-based international recession and Britain's entry into the Common Market. The latter weakened the Atlantic-facing west coast port; the former sent shock waves through many of the city's large, externally controlled firms. During the next decade a process of disinvestment, contraction, and closures sent the city's economy into a tailspin. During this period 60,000 jobs were lost and unemployment rose to 27%, twice the national average. However, the decline of the private sector in the city during this period was partly compensated for by the fact that, in employment terms, Liverpool had become primarily a public sector city—and this sector was growing. Expansion in central and local government, the health service, the university, the police, and the nationalized industries masked the contraction of the private sector.

However, when national government policy changed in the late 1970s, first under the impact of the Labour government's austerity program and later because of the Conservative government's commitment to reduce the public sector, the vulnerability of the city's economy was fully revealed. This had a major impact on the city's politics, in particular exaggerating conflict between its political parties about how Liverpool should respond to the combined effect of national policies and economic decline. The cuts in public sector expenditure especially encouraged the Labour party to the left and toward a direct confrontation with the national government. In turn, this reinforced the divide between business elites and local politicians.

If 1973 was a major turning point in Liverpool's economy, it was of even greater political importance. The restructuring of its electoral boundaries that year helped produce a major political upheaval as a minority Liberal party, which had previously held no more than a few seats on the city council, emerged to take control of it from the Labour and Conservative parties, which had previously dominated the city. But the Liberals never achieved an absolute majority on the council, and for the next decade the city was run by minority or coalition governments, which created constant political uncertainty. For much of the decade both partisan conflict over the minutiae of party politics and bitter competition between the parties to secure their volatile electoral support distracted attention from the larger issues of economic decline.

The Liberal administration of the 1970s set two major policy goals. The first was to dismantle the large municipal housing stock that had been built up by Labour and Conservative administrations in the 1950s and 1960s. The second was to reduce levels of rates (property taxes) to encourage the private sector to remain in the city. The Liberal administration's specific strategy to cope with the city's economic problems was a supply-led program, which, for example, provided serviced sites and advance factory units concentrated in particular areas to achieve maximum visible impact. The program also provided small grants, rent guarantees, and commercial advice to attract small firms, in particular, to invest in the city. However, the program was limited in its resources. It focused primarily on physical intervention in the economy. It was essentially grant-led and could not address the demand deficiencies of the local economy. In the scale of the city's economic problems, the program could make little impact, building about 150 factory units and creating or preserving less than 2,000 jobs at a time when tens of thousands were being lost in the city.

The Liberal administration's focus upon the private sector, combined with its efforts to reduce the scale of public sector housing and reduce local taxation, created intense hostility with the Labour party. This heightened partisanship, combined with both constant changes in administration and the construction and disintegration of coalitions, guaranteed that the city drifted through a crucial decade of economic decline with little clear response to it. At the same time, local business elites failed to contribute to the resolution of the difficulties by forging links with the city council. They were either contributing to the problem by retrenching and creating unemployment—or they were desperately attempting to protect their existing operations. During Liverpool's lost decade its economic failure was compounded by leadership failure.

1983-1987. URBAN REGENERATION UNDER LABOUR: MUNICIPAL SOCIALISM AND CONFRONTATION.

The economic and political failures of the 1970s laid the seeds of an even more traumatic political crisis in the city when Labour took control of the city in the 1980s. During the 1970s two important political realignments took place in Liverpool. In the first case, the Labour party gradually moved from being a minority party to inheriting the political dominance the city's class structure suggested it should have.

The Liberal party had seized control of the city at the beginning of the 1970s, but it actually lost ground to Labour throughout the decade, only surviving Labour's coming to power by squeezing the Conservative party out of the city. During the decade of the 1970s the politics of sectarianism were gradually replaced by the politics of class as the Conservatives inexorably lost their grip on working-class Protestant voters. By the beginning of the 1980s Labour was poised to construct the one-party state that dominates other industrial provincial cities in Britain. At the end of the decade it had become a reality.

The second political realignment in the city was that, as the electorate slowly turned to Labour, the party itself was moving from its traditional right-of-center ground to the far left. Indeed, during the late 1970s the party came under the control of the Militant Tendency, a Trotskyist faction that, although a minority in the party, exploited its organizational superiority to take control of the party leadership and shape its policies. When, in 1983, Labour won control of the city council with the first absolute majority in a decade, it had already decided to lead the city into a political confrontation with the Conservative government (Mulhearn, 1987).

Labour's victory ended the political uncertainty and lack of leadership; but the period of decisive leadership simply brought a new set of problems for the city. Labour was committed to reversing the priorities and policies of its Liberal predecessors. In particular, Labour had become increasingly frustrated with the Liberals' use of public funds to encourage the private sector while reducing support for public housing. In Labour's view the public sector had to be the engine of economic growth in a depressed city, and maximum resources should be devoted to preserving public sector employment in the city and rehabilitating and expanding the public housing stock in the most deprived areas of the city.

Labour's strategy encountered two major objections. One was that employment in the local authority or investment in housing was essentially investment in consumption, which did little to create long-term economic development. A more powerful argument was that, regardless of whether the strategy was desirable, it was unrealistic since it could not be funded. The Conservative government elected in 1979 was determined to limit the growth of the public sector and specifically was restricting revenue and capital expenditure by local authorities (Parkinson, 1987). This meant that Labour cities could not expand their work forces or sustain major house-building programs. However, despite these real constraints, the Labour administration, for the next four years, attempted to pursue both policies, bringing it into a major political confrontation with the Conservative government, the national Labour leadership, and finally the House of Lords.

From 1983, when Labour took control of the city council, until 1987, when the House of Lords finally disqualified most elected Labour councillors from office for their financial negligence, Liverpool's affairs were dominated by its highly publicized confrontation with the Conservative government over its revenue and capital budgets. Labour's fiscal argument and struggle was complex and the details have been rehearsed elsewhere. But it had a simple message: The local authority was, with 31,000 employees, the largest employer in a city that had in excess of 20% unemployment. Government cuts in the city's revenue and capital expenditure were further depressing the city's economy and creating fiscal stress for the council. Labour had an electoral mandate to resist the government's policy and use any political and financial tactics to sustain the level of economic demand in the city. For four years Labour attempted to force the Conservative government to provide more public funding for the city's employees and services by threatening to bankrupt the city if it were not given the extra resources. At the same time, it engaged in complex creative accountancy, which allowed it to finance a major public construction program despite central government efforts to restrict it (Parkinson, 1986a, 1986b).

Other Labour cities experienced similar resource cuts in the 1980s, and many reacted against them, particularly in a concerted national campaign against government policies in 1985 (Blunkett & Jackson, 1987). However, Liverpool had more extreme fiscal problems and a more extreme political reaction. Liverpool's case was exploited by the Militant Tendency, which wanted to use it to confront the Conservative government on a national level, either sowing public discontent or attracting more members through its highly publicized resistance. This determination eventually led the Liverpool Labour leaders to overplay their hand against the national government—and led to their downfall.

In 1984, after a bitter confrontation with the government and threats by the Labour council to bankrupt and bring civil unrest to the city, the government allocated marginal extra funding to the council to ease its fiscal problems. However, political exploitation of that victory by the Militant Tendency convinced the Conservative government it was counter-productive to help the city. In future years it resisted Labour's efforts to force extra money from it. Each year Labour refused either to cut its services or its employees or to increase its local taxes to pay for them. Instead, it insisted the government had a moral obligation to fund the council's revenue deficit. But each year the government refused to help, and the council balanced its books in a technically legitimate but financially imprudent way by transferring resources from its capital program to its revenue budget to pay part of its employees' salaries. Not only did the process merely postpone difficult budgetary decisions for future years by refusing to balance revenue and expenditure, but it also placed growing strain on the council's ability to fund its major public works capital program.

Until the Labour council was disqualified in 1987, it managed to sustain that program by borrowing almost £60 million from foreign banks, which would be repaid in future years. But as government tightened its rules on such schemes throughout the decade, the pressures on the council simply continued to grow. By the end of the decade, a new, more moderate Labour council was forced into increasingly desperate financial remedies just to sustain its revenue and capital program and balance its books.

The Labour council's primary effort to regenerate the local economy during the 1980s was its Urban Regeneration Strategy. Despite the scope of its title, the program was relatively limited. Essentially, it consisted of a public house-building program with environmental improvements and the creation of leisure facilities. Little attention was paid to other elements of an economic development strategy. For example, the growing potential of the city center as a retail, leisure, and tourism site was ignored. The potential of cultural industries was also ignored. Equally little effort was paid to supply-side measures such as training, the supply of capital, enterprise development, or the development of small and medium-size firms.

However, on its own terms, the strategy had a major impact on the city's physical environment. In 17 designated priority areas of the city, a major building and renewal program was undertaken, which in four years built more than 4,000 flats and houses and refurbished almost 8,000, built five leisure centers, and created three new public parks. Much of the urban fabric of the city was transformed and most of its late-Victorian or 1960s slums were either demolished or substantially modernized. The program, with the aid of creative accounting, extensive borrowing, and use of the receipts from the sale of the city's corporate estate, spent about £90 million a year at a time when the government was radically reducing public spending on housing.

Nevertheless, criticism of the program was extensive. Partly, the objections were to the housing philosophy of the council, especially the extensive municipalization of the strategy, which rejected alternative forms of tenure provided by housing cooperatives, housing associations, or private ownership. Also, the strategy concentrated resources upon a limited area of the city. Other areas deteriorated during this period, and the bulk of resources was allocated to capital works. Little money was set aside for maintenance, and the Labour council made little progress in improving the quality of its own labor force. The result was that many tenants received poor maintenance service, which undermined much of the value of the new building program.

More important in leadership terms, Labour's strategy alienated much of the private sector in the city. Major developers and construction interests naturally approved of the program, which provided them with so much work. But other commercial, retail, and industrial sectors found the concentration of resources almost entirely in the council's working-class heartland, and the dogmatic style with which it was being driven through the city, alien to their economic and political interests.

By 1986 relations between the political and business elites in Liverpool were at a nadir. The ideological and cultural differences between the two sides could not be more graphically underlined. But there seemed no obvious method of breaking that political impasse. Labour consistently won election victories with its strategy of municipal housing and political confrontation. The Militant leadership, which was driving the strategy along, although a minority of the party, had a powerful organizational grip on the party. This was reinforced by the fact that the party was dominated by blue-collar unions, many of whom were council employees and feared losing their jobs if the council adopted a more conciliatory strategy. And by this time an enormously powerful Labour politician had emerged to control both the city's financial affairs and its Urban Regeneration Strategy. He refused to make any concessions on the strategy to either external or internal party critics. His dominance emphasized another aspect of the weakness of leadership. The Labour party itself, with its workerist ideology and style and its cultural disdain for middle-class elements, produced very few alternative leaders who had the political or administrative experience to generate an alternative strategy around which internal critics in the party could mobilize.

1987-1990. PEACE BREAKS OUT IN LIVERPOOL: A NEW LEADERSHIP COALITION?

The political logjam in the city was finally broken by external forces—the national Labour leadership, the district auditor, and the courts. In the first place, the national Labour leadership, alarmed by the public relations damage Liverpool was doing to its national electoral chances, suspended the Liverpool Labour party in 1986, and subsequently expelled the leading members of the party for their membership in the Militant Tendency, which was a proscribed organization. In 1987 the House of Lords finally confirmed the judgment of a legal official, the district auditor, that, during its confrontation, the Labour councillors had failed to protect the financial interests of the council, and disqualified 47 of them from office. The subsequent reorganization of the Labour party by the national leadership, and the elections to replace the disqualified councillors, provided the opportunity for other groups to emerge to leadership positions in the party. Although a substantial minority of both the Labour party and the newly elected council remained members of the Militant Tendency or were supporters of their policies, the leadership of both the party and the city council passed into the hands of more moderate groups. They spent the next two years attempting to develop a way of staying faithful to the original Labour policy of sustaining municipal employment and building public housing, while finding an accommodation with the local private sector, potential external investors, and the national government.

The change in Labour's policies followed the elections in 1987 when, despite the expulsions and disqualifications, the party was returned to office with an even larger majority. But as a result of considerable internal maneuvering during the recruitment and selection process, the majority of the new councillors were closer to the political center or at least opposed to the tactics, if not all of the policies, of the Militant Tendency. During the first year of office a new party leader also emerged, a lawyer with considerable political experience who had the ability to balance the ideological interests in the party and reassure the local private sector that the policies of confrontation with them and government would be abandoned. Such Labour guarantees of Labour moderation were virtually inevitable. It had become clear to all sectors in Liverpool that the politics of confrontation had not only failed to extract extra government resources for the city, but had also lost its local and national allies. Equally, it was becoming difficult for Labour to sustain its predecessors' massive capital works program since, between 1983 and 1987, that program had anticipated and consumed future resources, leaving the new council to cope with the financial consequences.

Labour's new strategy did not replace the goals of maintaining local authority employment and sustaining a housing program. But it generated new priorities—in particular, a focus upon the new, as opposed to traditional, functions of the city, which emphasized their consumption as much as their production functions. Whereas the earlier Labour strategy had concentrated on municipal services in the most deprived areas of the city, the new strategy emphasized the potential of the city center in terms of retail, leisure, tourism, and commercial development. The shift partly reflected the fact that the city did not have the resources to continue the massive house-building program. It also reflected a realization that although the council could exercise little leverage over the operation of the private market, it did exercise a degree of control over the city center environment, which provided more than 40% of the jobs in the city. This realization was encouraged by the fact that the city's waterfront, which was no longer under its control, had been substantially rehabilitated by Merseyside Development Corporation and was beginning to serve as a focus for a growing leisure and tourism sector (Parkinson & Evans, in press).

During the following two years, a stream of policy documents came from the city council advocating the need to diversify the economic base of the city and, in particular, the need to regenerate the city center as the center of a regional market of 3 million people. The city's dilemma was that such proposals needed resources and cooperation between a variety of sectors over which it had little direct control. The council was essentially setting the agenda for the regeneration of the city without necessarily being able to mobilize the necessary financial, administrative, and human resources to implement it. Nevertheless, the council did make substantial changes in the one obvious source of money available to it, the Urban Programme that was directly provided by the government. The previous Labour administration had used the program of approximately £20 million a year to support its housing and environmental programs. The new administration used this money differently—to pedestrianize substantial parts of the city retail zone; to create a park linking the city center to the waterfront; and to provide funds for the refurbishment of the city's theaters.

The funds were also used to provide training grants to local businesses that employed apprentices, to develop a security improvement scheme in inner-city sites, and to publicize the incentives available for potential investors in the city. The sums of money involved were hardly crucial. The importance was that the initiatives reflected the priorities of the Conservative government's urban strategy, and many of them actively involved the participation of business groups like the Chamber of Commerce, the City Stores Committee, as well as the Merseyside Development Corporation. Essentially, the council used its limited financial resources to send a signal to the government, the local private sector, and potential investors that the era of confrontation was over and the council recognized the need to form a partnership with them.

A series of related small initiatives encouraged this view. The city commissioned a variety of consultancy studies to examine the tourist potential of the city, the problem of the city's image and city marketing, the economic potential of the design industry, and, most significantly, the merits of creating in the city a partnership between the public and private sectors to guide economic development. Again, the significance of such studies was clearly not that major policy consequences flowed directly from them, but that they demonstrated different intentions from their predecessors. By 1990 Liverpool was at least beginning to follow the path that Glasgow and Sheffield had adopted earlier in the decade. Whether that change in style would lead to sustained economic regeneration would clearly be affected by many other factors, only some of which were under the city's control.

CONSTRAINTS ON REGENERATION

The least-forgiving constraint on the city at the beginning of the 1990s remains its structural economic circumstances. Liverpool has experienced either job loss or slower growth in each major sector of the economy and a consistent under-performance, compared with national trends during the past 25 years. It has lost 61% of its manufacturing jobs since 1961, which provide only one in five jobs in the city and which remain heavily concentrated in a small number of large firms. The city lags behind the national average in the growth of the service sector, which has suffered from both the decline of the port and reduced demand from business and individuals. One-quarter of the city's population of working age lacks a permanent job. More than 30,000 Liverpool residents have been out of work for more than one year, and more than 16,000 for more than three years. The work force remains relatively underqualified and unskilled and poses a major task of long-term investment in human capital (Liverpool City Council, 1989; MIDO, 1989).

The city council itself also remains as a constraint on regeneration. As the largest employer, the largest service provider, and, in many respects, the political leader of the region, the council continues to face major problems. The quality of its services, which deteriorated during the 1970s because of poor management, remains poor. The council has developed an internal culture that is fragmented and negative, which prevents major administrative reorganization. Many of its services—housing maintenance, environmental services, the schools—are of low quality. And the council suffers from a lack of clear administrative leadership. In 1988 the Chief Executive of the city resigned because of his inability to move the administrative machinery, and the city has still been unable to recruit a senior figure to replace him.

In addition, the council faces major financial problems that make it politically difficult to rationalize and administratively difficult to innovate. Although it has abandoned the high-profile confrontation with the government over its fiscal affairs, the Labour council remains committed to local authority employment and at least the maintenance of the capital works program. However, a decade of creative accountancy, in which the long-term capital resources of the city were used to support the city's short-term revenue expenditure, is making both increasingly difficult to achieve. By the end of 1989 the city treasurer indicated that the council could only sustain both of those commitments if it sold to the private sector more than £60 million of the city's prime commercial and industrial assets. Regardless of the merits of such short-term asset stripping, it is becoming doubtful whether such sales could be achieved.

The regeneration of the city in the late 1980s is also constrained by a legacy of industrial and political militancy, which, even if no longer entirely accurate, leads potential investors to regard the city as a risky location. The combination of difficult labor relations in the port and the car industry in the 1960s and 1970s, and the political confrontation with the government in the 1980s, means that the city continues to have a major image problem that will require a long-term campaign to redress. There also remains an internal image problem. Two decades of economic failure, compounded by political failure and self-destruction, have bred a degree of cynicism in the city's public life. There is clearly a cultural dimension to the city's failure that goes beyond the statistics of economic decline (Parkinson, 1988).

At the end of the 1980s, however, there are some indications that Liverpool has at least reached the bottom of its economic fall, and some signs that an economic recovery could be exploited. The manufacturing industry that remains has weathered the recession of the mid-1890s restructured and reinvested. Automobiles, engineering, and printing have shed jobs but become more efficient to form the core of a modern manufacturing industry. Liverpool has also witnessed some growth in the service sector. Approximately 10,000 jobs were created in financial services in the region during the decade. If this lags behind other regional competitors, it nevertheless confirms that there could be some growth in that sector.

There has been an expansion into some high-technology industries, especially in the small but successful Wavertree Technology Park. The port itself has also rationalized during the 1970s to become an efficient and profitable enterprise, even if its work force has declined from a postwar peak of more than 40,000 to less than 3,000 in the late 1980s. The waterfront is being renovated by the Merseyside Development Corporation, and tourism, leisure, retail, and light industry activities are beginning to flourish. Retailing in the city center is growing in keeping with the national trends. There has been a limited amount of office-building and rehabilitation in the city center, and office rentals are beginning to rise. Although they lag behind other more successful cities, the indications are that there is at least some demand. The residential market was also transformed in the late 1980s as property prices in the city virtually doubled between 1987 and 1989.

THE REMAINING DIFFICULTIES

At the end of the 1980s Liverpool appeared to be at a turning point. The economic collapse of the early 1980s had been arrested. There was some growth taking place in some modern sectors of the economy that had growth potential. The political complexion of the city had changed, and many of the internal divisions of the recent past seemed reduced. Cooperation rather than conflict characterized relations among the city council, the private sector, the Merseyside Development Corporation, and government officials—if not always politicians. There was some agreement among many of the major institutions as to which sectors of the economy had development potential and should be exploited. There was a clear recognition among elites of the problems that recent political confusion and conflicts had created for the investment potential of the city.

Nevertheless, there remained substantial difficulties. The economic recovery of the city was fragile and vulnerable to downturns in the national or regional economy. The economic growth that had occurred was uneven, concentrated in particular sectors of the economy and areas of the city. Substantial unemployment and physical dereliction in many parts of the city posed continuing economic and potential political problems. The economic and political exclusion of the Black community from the mainstream life of the city remained a divisive and potentially explosive issue (Gifford, 1989).

At the end of the 1980s urban entrepreneurialism had replaced municipal socialism, the favored strategies of provincial cities seeking new economic roles in 1990s Britain. A number of Labour cities, including Sheffield and Glasgow, were beginning to pursue that path and were having some apparent success. Their experience indicated that the policies and actions of leaders could affect the development path of the city. It at least suggested that positive action was preferable to blind acquiescence in the face of economic change, even if one could not control it. Liverpool finally had chosen to imitate the model. But its range of structural economic, environmental, political, and cultural problems suggests the city would remain one of the most testing cases for urban regeneration. Liverpool's leadership problems during the 1970s and 1980s suggest that political failure can exaggerate economic failure.

REFERENCES

Audit Commission. (1988). *A review of Liverpool's services.* London.

Baxter, R. (1972). The working class and Labour politics. *Political Studies, 20* (1).

Blunkett, D., & Jackson, K. (1987). *Democracy in crisis: The town halls respond.* London: Hogarth Press.

Cornfoot, T. (1982). The economy of Merseyside 1945-1982: Quickening decline or post-industrial change. In W. T. S. Gould & A. J. Hodgkins. *Resources for Merseyside.* Liverpool: University of Liverpool Press.

Crick, M. (1986). *The March of militant.* London: Faber & Faber.

Lord Gifford. (1989). *Loosen the shackles.* London: Karia Press.

Harding, A. (In press). Public-private partnerships in urban regeneration. In M. Campbell (Ed.). *Local economic policy*. London: Cassell.

Hindess, B. (1971). *The decline of working class politics*. Paladin.

Lane, T. (1987). *Gateway to empire*. Lawrence Wishart.

Liverpool City Council. (1989). *Liverpool City Council—prospects of the Liverpool economy to 2000*.

Lloyd, P. (1979). The components of industrial change for Merseyside inner area: 1966-75. *Urban Studies, 16* (1).

Lloyd, P., & Dicken, P. (1978). Inner metropolitan industrial change, enterprise structures and policy issues: A case study of Manchester and Liverpool. *Regional Studies, 12*.

Markusen, A. R. (1989). Industrial restructuring and regional politics. In R. A. Beauregard, *Urban affairs annual review: Vol. 34. Economic restructuring and political response* (pp. 115-148). Newbury Park, CA: Sage.

Merseyside Integrated Development Operation—1988-1992. (1988). Liverpool City Council.

Parkinson, M. (1985). *Liverpool on the brink*. Policy Journals, Hermitage.

Parkinson, M. (1986a). Creative accounting and financial ingenuity in local government. *Public Money* (5), 27-32.

Parkinson, M. (1986b). Decision-making by Liverpool city council. In *The Widdicombe inquiry into the conduct of local authority business: Research Vol. No. 4. Aspects of local democracy*. London: HMSO.

Parkinson, M. (Ed.). (1987). *Reshaping local government*. Policy Journals, Hermitage.

Parkinson, M. (1988). Liverpool's fiscal crisis: An anatomy of failure. In M. Parkinson, B. Foley, & D. Judd (Eds.), *Regenerating the cities: The U.K. crisis and the U.S. experience*. Manchester: Manchester University Press.

Parkinson, M. (1990). Political responses to urban restructuring: The British experience under Thatcherism. In J. Logan & T. Swanstom. *Beyond the city limits*. London: Temple University Press.

Parkinson, M., & Evans, R. (1990). Urban development corporations. In M. Campbell (Ed.), *Local economic policy*. London: Cassell.

Taafe, P., & Mulhearn, T. (1988). *Liverpool—a city that fought back*. Fortress.

Waller, P. J. (1968). *Democracy and the rise of sectarianism*. Oxford: Oxford University Press.

13

Recasting Urban Leadership in Buffalo

DAVID C. PERRY

REGENERATION AND RUSTBELT CITIES

By many experts' estimation the urban "rustbelt" of the United States is beginning to shine again. The financial press is replete with stories suggesting that the conditions of global restructuring have turned from decline to regeneration in the urban North, the product of a reconfigured regional economy which looks, in sectoral and spatial terms, very different from that of the early twentieth century (Beauregard, 1989; Bingham & Eberts, 1990; Markusen, Naponen, & Driessen, 1989).

Just as the generative *image* and sectoral *composition* of urban economies have shifted (Judd & Parkinson, 1988) so too have political relations been recast. In one region after another, private sector elites are exerting highly visible and decidedly more organized roles than they did, in some cases, just a few years ago. Whether they are acting more deliberately public than in the past or are actually practicing a new role, the fact is that business leaders are giving the appearance of taking charge of the urban centers in which they live. While the organization of capital in each of these cities has been different, there appears to be a common theme justifying the new role of business: namely, that government will not work without direct capital intervention. Speaking before the National Council of Economic Development officers, Cleveland Mayor George Voinovich explained: "Good government today does not exist without the help of an active, supportive, private sector. By private sector, I don't mean some faceless, distant actor. I mean an active partner. Successful government is a productive *public-private partnership*" (Voinovich, 1988). It should come as no surprise to learn that Mayor Voinovich was the candidate chosen by the Cleveland Tomorrow group to lead the city out of default to economic recovery—to create the "miracle on the Cayuhoga" (Magnate, 1989).

For some in the financial community (*Wall Street Journal,* 1988) a similar miracle is underway in Buffalo as well. A group of business leaders, motivated by Cleveland Tomorrow, has been formed—the "Buffalo 18." While the intervention of capital in the public realm is as old as the urban regime itself (Smith, 1989), the emergence of these corporate groups in visible and powerful positions in the regeneration process represents a new and ostensibly different version of private sector action: a *recasting of urban leadership*.

The term "recasting" is used deliberately—in an almost theatrical sense—in order to suggest that the pattern of direct capital intervention in the economic development/regeneration process represents not so much a new "play" of urban political economic relations as it does a redefinition of the "roles," the "players," and their place in the political economy of regeneration. In the past, business leaders did not participate in urban governance because they did not have to in order to maintain production in the firm and investment in the region. In fact, for much of the past few decades, the role of many regional CEOs was to close down firms and disinvest in the region. In today's era of regeneration, it can be hypothesized that the logic of production and entrepreneurial investment may dictate a certain reversal of this pattern—necessitating a new civic role for the CEO. This chapter attempts to detail the beginning stages in such a "new casting" of CEO participation in the regeneration process of Buffalo, New York.

BUFFALO: FROM BUST TO "BOOM" TOWN?

Just a few years ago, the slogan on one of the most popular T-shirts in Buffalo conveyed how many of the region's residents, from the street level to the board room, viewed the economic future. It read: "Buffalo: City of No Illusions." For most of the preceding two decades, the region had experienced massive disinvestment with the virtual elimination of the steel industry, the visciation of the auto industry, and the decline of its once-vaunted industrial base from a position of superiority to near economic emasculation. Where manufacturing had accounted for almost one out of every two jobs in the 1950s, it accounted for less than one out of five jobs in 1988 (*Wall Street Journal,* 1989).

At the same time growth has occurred in Buffalo's service sector—those economic activities, including financial and business services, retail and wholesale activities, and health and educational services. Even here growth remained an illusion: While it generated many new jobs (Rudnick, 1989), they were not enough to make the region nationally competitive as a center of finance, education, producer services, or other new base economies. Today, even with substantial increases in employment in the Buffalo area since the recession, the region is still far from competitive, relative to the rest of the country.

The demographic transformation of the region has been every bit as dramatic. Between 1970 and 1980, the region lost more that 5% of its population—almost 95,000 people. It is estimated that another 38,000 people left in the ensuing five years. Just as the economy disintegrated at its base and has not been reconstituted competitively through another constellation of industrial sectors—so, too, has the productive base of the population started to erode. Using present population trends as a base, it is projected that the region could well lose more than 86,000 people between the ages of 15 and 24. On the other hand, the region stands to gain the most growth in the population in the age bracket of 65 and over. By the year 2000, it is estimated, those in this category will increase by almost 200,000—an increase of 30%—which represents, far and away, the most dramatic increase in population in the region. By the year 2000, the smallest age cohort will be the 15-24 group and the largest age group will be that over the age of 65 (Perry, 1988).

The central city of Buffalo continues to house a substantial share of the region's minority population. Between 1970 and 1980, more than 100,000 whites left the central city and the Black population continued to stay in the city, with very little evidence of outward mobility. In fact, the city's Black population is now almost 30% of the population. Together with Hispanics, they now account for the largest group using the city's public schools, as well as receiving welfare and transfer payment assistance (Kraushaar & Perry, 1990).

Finally, the economy, transformed as it is, is not providing a per capita income that lifts the population above the national average. While per capita income in the Buffalo SMSA increased between 1970 and 1985 by 18.3%, from $8,544 to $10,016, this was still below the national per capita income figure in 1985 of $10,132. Somewhat perversely, one public official now counts the sagging income level of Buffalo workers as part of the "cheap labor base" of the "rebounding business climate" (Rudnick, 1989).

This leader is not alone in the attempt to fashion a strategy of economic renewal out of Buffalo's fragile economic structure. As in other cities, the role of capital in the public execution of economic change is being recast—in search of new advantages in a time of ongoing economic change. As one corporate leader put it, in the early 1980s:

> It became clear to many members of the business community that our self-interest, our very economic as well as personal self-interest, was actually influenced by what happened to the Buffalo economy. We really had only two choices—get involved or leave. As we looked around the city and region we were appalled—the region was not only in decline economically—it was out of control. It might sound awful, but in 1983 we set out to change this—to get the city and the region back under control (Interview, 1989).

The Buffalo region in the early 1980s was becoming a place where corporate leadership, for whatever reason, could not personally continue to live. It was out of "control."

FROM GOLDEN AGE TO CITY OF NO ILLUSIONS: PATTERNS OF URBAN LEADERSHIP.

One century earlier, as the City of Buffalo entered its 50th year, it was lauded as the premier city of the continental interior, with almost 1,200 manufacturing firms and a population of 170,000. There were 15 railroads in Buffalo, which caused one leader to suggest that "we are the railroad center of the world." Buffalo was midway through what historians would later call its Golden Age—a period stretching from the end of the Civil War to the early decades of the twentieth century.

At this time economic success had produced a certain smugness: "Buffalonians contentedly went . . . [about] . . . their jobs, which were plentiful. Businessmen, all seemingly possessed of the Midas touch, went on with the business of drenching themselves in dollars. . . . [They came to Buffalo] in droves and they stayed and prospered . . . "(Brown & Watson, 1981). They joined some of the most visible business organizations of the day. For example, not a member of the City's Merchants Exchange could be found "who did not rightfully think of himself as a capitalist of the highest order and wear the trappings of capitalism with all the pride he could muster" (Brown & Watson, 1981). Or, as the nineteenth-century Buffalo attorney, Eben Carlton Sprague, put it: "Buffalo is not a snobbish city. There is no city where solid wealth is more sincerely respected."

In the early years of this Golden Age, it was Buffalo-born companies that headed many of the nation's major economic sectors, including lumber, meat-packing, pottery and clay goods, steel and iron works, breweries, soap, chemicals, and even autos. In fact, at one time in the early twentieth century, 28 different automobiles were manufactured in Buffalo, including the Pierce Arrow.

All this activity and wealth did not seem to generate a clear spillover of private sector leadership in the public sector. In fact, the same prominent lawyer who had declared that Buffalo was not a snobbish city also saw in this pattern of capital accumulation an inherent selfishness. "The expansion of . . . this wealth in due time became a matter of concern to Eben Carlton Sprague. [In 1884] Sprague issued this dire warning to a captive audience of Buffalo's wealthiest: It was wealth without conscience that sowed the seed of the French revolution and drove its possessors into exile and to the guillotine. Spread it around, he advised the Buffalo rich . . . " (Brown & Watson, 1981).

After the Sprague admonition, some members of the private sector of Buffalo would indeed "spread it around," starting the philharmonic, the Albright Museum, the historical society, and the like. Others would become part of the nascent business reform movement, which attempted to throw some of the rascals out of city hall. Still other Buffalo enterpreneurs would practice a very different form of "spreading" their business, if not their wealth, "around." The rapid growth of the Western New York region made Buffalo, by the turn of the century, one of the most attractive centers of production and consumption in the industrial era: Soon names like Morgan, Vanderbilt, Mills, Lehman, and Rothschild began to appear on the ownership, stockholder, and partner lists of local firms. Midway through the Golden Age, many local capitalists began a pattern of "spreading it around" that included neither Eben Sprague's call to philanthropy nor a reformist's zeal for change in the name of the public interest. Buffalo's leading capitalists had entered a new round of wealth generation, built not upon the production of industry in Buffalo but upon its sale to outside investors (Goldman, 1983). The result: By the 1930s, "the manufacturing base of Buffalo was essentially owned by outsiders" (Goldman, 1983). The Golden Age had ended.

The political arena remained as separated from the business elites as ever—the bailiwick of a succession of new ethnic groups and the product of countless party upheavals in both the Democratic and Republican parties. In fact, with the families of Buffalo's early entrepreneurs voting with their "stock options," the political leadership of the region came to be centered, both symbolically and really, in the office of the mayor. Historian Scott Eberle concludes that politics in Buffalo became a revolving door through which one ethnic group after another marched on its way to the mayor's office and temporary control over city hall.

Few regions more accurately reflected the boom-bust cycle of the Depression and wartime industrialization than Western New York. The area had become less a center of entrepreneurial, homegrown manufacturing and more a region captured by the organizational logic of mass-production industrialization, wherein manufacturing would be carried on in firms built by the twin dynamics of the war economy and the restructuring of industries through mergers and consolidations. In the earliest stages of postwar growth of the auto and steel industries, Buffalo suffered an event which, in retrospect, would presage the region's economic decline: the close of the Curtiss Wright Aircraft Plant and the loss of 40,000 jobs. A decade later the region's DuPont plant would move away and the new divisions of the major chemical giants—Allied Chemical and Hooker—would not open in their old home base of Western New York. In less than another 10 years, the American Shipbuilders would close and five of the nation's largest grainmilling operations would shut down.

By 1970 manufacturing employment in the region was in a state of absolute decline, while such traditionally high-growth sectors as metalworkers generated only 11 new jobs in a decade. Steel, auto, and metalworking were all industries in competitive and productive trouble, and the decision rules of capital during this period argued for plant closings and layoffs throughout Western New York. For most local business leaders, they were not deserting their home town, they were just responding to the changes in global competition that forced the shutdowns and disinvestment in the region (Perry, 1987).

By the early 1980s the pattern of long-term local neglect and political-economic powerlessness was especially apparent to a new group of outsiders who had moved to Buffalo to head some of the region's largest industries and banks. The members of this new group were highly entrepreneurial and they were used to living in communities where economic growth was a watchword (Lipsey, 1989). As one person put it much later, "we were not going to be known as a group who had moved to a backwater. We had not come to Buffalo to fail." There was Stanford Lipsey, the Warren Buffett-selected publisher of the *Buffalo Evening News*—the daily newspaper. And Ross Kensie, the controversial and highly visible banker brought in to head Goldome Bank, at the time the eighth-largest savings and loan in the United States. Robert Wilmers was head of a New York City group that bought the local M. and T. Bank. Another member of the new leadership was outsider Dr. Steven Sample, the new president of the State University of New York at Buffalo.

THE RECASTING OF URBAN LEADERSHIP IN BUFFALO

The infusion of outside leadership came about at the same time that the regional economy reached its nadir. If David Campbell of the Computer Task Group worried that things just didn't work anymore, Stanford Lipsey, fresh from the Sunbelt, could not believe that the region could be any worse off:

"Not only was the economy and the political structure in disarray, no one was even talking to each other about it. It was as if every one was living in their own world, protecting their world from the failure of the economy while assuming the worst for the region without consequence to themselves" (Lipsey, 1989).

As Lipsey made the rounds in his capacity as the new publisher of the paper, the lack of communication and the sense of benign acceptance of the failures of the region disturbed him greatly. In late 1983, with the help of his editor, he inaugurated a series of breakfasts with 20 of the most influential CEOs in the region. He then organized a smaller group: banker Robert Wilmers, CEO David Campbell, Norstar Bank executive Eugene Mann, and State University President Steven Sample. The subject of the meetings was always the same: "Buffalo is in trouble—who can we identify to help?" (Lipsey, 1989).

THE POLITICAL REORGANIZATION OF LOCAL CAPITAL

In the first meetings, Lipsey was a key player. Everyone was willing to talk to him because he was the publisher of the paper. Commented David Campbell:

> He was as neutral as you can get and still be a member of the business community . . . In those early meetings we made up lists of who we wanted to join us. The major decision rules were simple and clear. We wanted the group to be as small as possible and as powerful as possible. That is we wanted to put together the smallest number of people it would take to make sure we had the key people to guarantee that any idea we had would succeed. The first ten members were easy. Then we just went around the region to make sure we had all the rest of the right names—the traditionally most powerful people like the Jacobs, the Knoxes and the Riches (regular members on most lists of the richest families in the world). With these eighteen we seemed to have created a "winning hand" [Campbell, 1989, p. 4].

The only members asked to join were CEOs. Banker Ross Kensie said:

> It was clear from the beginning that the real power of the group lay with the ability of its members to deliver on whatever they promised at the time they promised it. Only CEOs can attend a meeting and under no circumstance can a member send a representative. If we decide it is necessary to raise $100,000 for the Philharmonic, we don't want to have some members say it's a good idea

TABLE 13.1
The Buffalo Eighteen: Membership, By Sector

Manufacturing
Durable Goods
Toys—Bruce Sampsell, *Fisher-Price*
Nondurable Goods
Pharmaceuticals—Wilferd Larson, *Westwood Pharmaceuticals*
Food Products—Robert Rich Jr., *Rich Products*

Services
Transportation
Water Transport—D. Ward Fuller, *American Steamship*
Whole/Retail
Whole/Retail—Jeremy Jacobs, *Delaware North*
Food Stores—Charles Barcelona, *Peter J. Schmitt*
Savino Nanula, *Niagara Frontier Services*
General Merchandise —**Robert Adams, *AMA's* retail stores

Producer Services
Elect/Gas—Louis Reif, *National Fuel Gas*
Banking—*Eugene Mann, *Norstar Bank*
Robert Wilmers, *M & T Bank*
Ross Kensie, *Goldome Bank*
Northup Knox, *Marine Midland Bank*
Security Brokers—Seymour Knox, *Kidder Peabody*
Bus Services—David Campbell, *Computer Task Group*
Real Estate—Paul Snyder, *Snyder Assoc.*

Social Services
Nursing Services—Frank McGuire, *Nursing Homes*
Education—Steven Sample, *SUNY, Buffalo*
Miscellaneous—Stanford Lipsey, *Buffalo News*

* No longer a member
** New member
SOURCE: Buffalo Evening News Files 8/10/89.

but they have to go back to their bosses or board. The strength of the group rests with its ability to act without having to check with anyone [Kensie, 1989].

By 1984, the final list of participants in the "Buffalo 18" appeared to be set—17 CEOs and the president of the region's largest public university (and not coincidentally, the region's largest employer). If the members on the list constituted the "new economic leadership" of Buffalo, then it also represented the sectoral shift in the economy as well. Of the 18 members, 15 were from the service sector—representing 11 different services (See Table 13.1).

Only three were from the manufacturing sector. It is estimated that, together, the group's firms employ more than 40,000 people or in excess of 10% of the employment base of the Buffalo area.

Fully half of this new group was not from Buffalo. It is this group of outsiders that is most often credited with the new organization of private leadership (Anzalone, Lipsey, & Sullivan, 1989). As one Buffalo 18 member put it: "If Buffalo had not had this new blood, nothing would have happened. It would be dead as in the past. All the major local industry had been sold and Buffalo did not have enough interested sons and daughters to carry on" (Kinsie, 1989). Historian Mark Goldman reflected on this "surprising" intervention of the long-dormant private sector into public activities this way: "The only game in town had been the patronage of the public sector and the labor unions. As industry died, the politicians became even more powerful because the public sector became the only place where lots of jobs were and where public subsidies came from" (Goldman, 1989).

For almost everyone, the climate of decline was being serviced by a vacuum of leadership (See interviews with Anzalone, Sullivan, Eberle, Savage, Rudnick, Goldman, Tobe, & Fuller, 1989). While there had been a high level of political activity in the region, it was incredibly isolated from the cycles of the economy. As community development, housing, and block grant funds dwindled as part of two decades of fiscal and programmatic retrenchment at the federal level, the influence of local political leaders dwindled as well (Eberle, 1989; Kossy, 1989; Tobe, 1989).

"This region cries out for leadership," said D. Ward Fuller, CEO of American Steamship Inc. University President Steven Sample echoed Fuller: "A public official or a public group can be a catalyst, but without strong, devoted, private sector leadership, it would be hard to see an economic renaissance anywhere. Conversely with that leadership it's hard to believe that we won't experience it" (Sullivan, 1986). These were strong words from a leadership sector that most Buffalonians had come to expect little, if anything, from in the form of public pronouncements, much less public action.

Group members may have been invoking the call for leadership, but they did not appear to have a clear notion of what this meant. Soon after being chosen by Lipsey's small "kitchen cabinet," the Buffalo 18 met for dinner. The only outside guests were representatives of the Cleveland branch of the consulting firm of McKinsey and Co. This was the firm that had been employed by a similar "cabal of CEOs" in Cleveland (*Fortune,* 1988) to design a program for that city's recovery. In Cleveland, the McKinsey group had suggested the creation of a formal organization to act as a vehicle for CEO intervention into the process of reerecting the region's economy. The result was "Cleveland Tomorrow." Stan Lipsey, among others, had been impressed with the first few years of the Cleveland experiment. It seemed only logical that one of the first efforts at setting an agenda and structure for the Buffalo group would be to invite the McKinsey consultants to suggest an organization for Buffalo. Banker Wilmers suggested just that.

After the representatives of McKinsey finished their presentation, they were asked to leave the room. The first decision of the nascent "group of 18" was to ask the McKinsey consultants to leave the city. It was clear that the proposals of McKinsey were, at best, premature. It was also clear that the best thing about the group was the fact that it was meeting. The meeting itself was somewhat historic. "Up to this time, the members of the business community of Buffalo not only didn't deal with the public sector, they didn't even talk to each other," said New York City transplant Robert Wilmers. Sounding very much like a latter-day Eben Sprague, he observed that the business leaders of Buffalo "were arrogant, petty and isolated" (Wilmers, 1989).

Therefore, the organizational basis for the group became, and remains, very simple—*communication.* Beyond this, a decided informality characterizes the group's structure. It has no name, no charter, no staff—not even any stationery. Its members never take a vote. One member, food store magnate Charles Barcelona, says the group is "nothing more than a forum which can have some influence" (Sullivan, 1986). But banker Ross Kensie sees this studied informality differently—"we put all the real decision makers n Buffalo around one table with no formal agenda other than getting the region off the dime. This lack of a public agenda drives the politicians nuts."

But the question remained: Why did these busy, powerful CEOs meet? Were they really the altruistic alternative to the old selfishness identified a century ago by Eben Sprague? For "18" founder David Campbell, the answer was clearly no: "make no mistake about it," he said, the leadership of the Buffalo economy joined in this process of regeneration out of "self-interest—business-related self-interest. Of the outsiders who joined us, no one really wanted to be associated with a community which was failing. For all of us, outsiders and lifetime residents, it was clear that the health of the region was a part of the health of our businesses."

If the organizing principle was communication and the structural form was deliberately informal meetings, the strategy of the group was just as deliberately ad hoc. The members decided early on that there would be no group intervention into political elections. Unlike business leaders in Cleveland and New York City, they eschewed the role of kingmaker—refusing to select or endorse any candidate for political office. When they wanted to make a point with a political leader, they employed a variety of tactics to make their collective will known to public officials. For example, one of their first actions was to sign a manifesto of intent to influence the economic development of the Buffalo area. The document was sent to Governor Mario Cuomo, with a request that he meet with them. Cuomo flew to Buffalo and, impressed "by the roster and the by the first meeting, Cuomo initiated a second get-together" (Sullivan, 1986).

Individually, they can be just as impressive—one of the members of the 18 is a close advisor of the mayor of Buffalo, James Griffin, and another member of the 18 is not only past-president of the powerful Western region of the New York State Business Council but also plays occasional one-on-one basketball games with the governor. One public official went with this athletic group member to see the governor one day, expecting little in the way of action on regional issues. The formal meeting with the governor was not productive, but the basketball game brought the results the official wanted.

Some members of the group lament the lack of direct intervention into electoral politics, but with disagreement on both the act of intervention and on candidates, the issue is moot. What are not moot are other forms of direct intervention into the process of regional economic regeneration. If there is one critique that consistently appears in the private sector's list of things that are wrong with Buffalo, it is the lack of good management and efficiency. And if there is one thing that all the members of the 18 believe they can contribute to the region's regeneration process, it is influential advise on the sound and efficient management of the governance of economic development.

To this end the Group of 18 has *directly* intervened in the selection of many public and not-for-profit leaders of economic development agencies, cultural organizations, educational institutions, and health and related research and development organizations. They have also employed a version of the interlocking directorate model and, where there is such an entity, have placed one or more of their members on the governing board of every one of these agencies. In a way, they have sought to become the region's Board of Boards—the center of technical, managerial, and decision-making power in both the public and private agencies directly concerned with regional regeneration. They have not sought to participate in the daily activities of regeneration, but merely to guarantee that the process *will* occur in a "rational and efficient way. A way that is a far cry from the way the political and economic communities have acted for years" (Kossy, 1989).

The substantive agenda of the Group of 18 is less one of its own creation and more that of the special agencies it has started to influence. In the past five years this agenda has grown to include: the development of what is described as the largest urban waterfront in the United States; state-generated economic development programs; the economic marketing of the region; local and regional economic development loans and grants to private industry; central city development; public infrastructure and transportation; regional economic development strategy and planning; advanced research and development; higher education; public education; and cultural development and quality of life.

It is clear that the group, no matter how informal, is having an impact. Its members take direct or indirect credit for hiring a large number of the agency heads specifically assigned to the issues of regional regeneration. This process is "really an easy one for most of us," says one Buffalo 18 member, "we have scoured the country to find the very best person for many of these jobs. In other cases we have had members in the executive position on the Board of Directors who approved the appointment of chief officer. This is just what we do in our own companies—search and hire the very best. Its just good business." Another business practice familiar to all "18" members is participation on governing boards: In the first four years of its existence, the group placed one of its members in a leadership position on every public board of directors except the Niagara Frontier Transportation Authority.

THE ORGANIZATIONS OF REGENERATION

Moving the economic development function off-line and into the bailiwick of *private*-public partnerships (Feagin & Smith, 1987) and other forms of special purpose governments (Leigland, 1989) has become a common feature of the regeneration process. As the previous section indicates, this has also been the case in Buffalo. The regeneration process is only partially captured by the traditional agencies of county and city government. The reason for the central importance of the city in development rests, in large part, with the particular abilities of the present mayor, James "Jimmy" Griffin. He succeeds in the political world of economic change by subscribing to a policy dedicated to getting the "first shovel in the ground." For him regeneration is not planning, it is action.

Among most of the Buffalo 18, the mayor is admired for his action and criticized as yet another example of the "fractious and insensitive and inefficient politics which has long plagued the region" (Lipsey, 1989). Even one of his key supporters among the 18 says that the mayor and "his style of government have been around too long. The mayor represents the end of an era of unenlightened leadership" (Kinsie, 1989). Enlightened or not, the mayor remains the single most important political leader in the region's regeneration process. He is the one political official who the new business leadership of the Buffalo 18 cannot influence. He has carefully centralized his authority as the "CEO" of city development: personally overseeing the entire political process of land rationalization, appointments, and funding. In fact, only after four years did he allow his economic development officer, Charles Rosenow, to begin to attend economic development meetings with other "18" appointed officers, and, in the words of one of his peers, "The mayor keeps Chuck on a very tight leash."

The rest of the economic development process of the region has been far more accessible to the influence of the group. In a way, members of the Buffalo 18 have come to view the structuring of urban regeneration activities like the reorganization of a corporation into "product divisions." It has been easier to relate to activities like public transportation, industrial revitalization, education, and cultural programs when they are broken into special purpose governments or not-for-profit private development groups.

In Buffalo the proliferation of such special purpose organizations has become a hallmark of the recasting of its urban leadership. Since its inception in 1984, the "Group of 18" has become very active in the reorganization and policies of old-line "product divisions," such as the Chamber of Commerce and the Buffalo Philharmonic. More recently the group has lent its support to the development and reorganization of the area's industrial development authorities, public works authorities, privately sponsored development foundations, and not-for-profit development groups.

The role of the Buffalo 18 has been one of support and reorganization. The group has not sought to initiate new units of government or new private foundations or development groups. In the past five years the members have placed themselves in positions of influence either as board members or as advisors to key public officials on a select group of organizations and public agencies in order to:

(1) market the region to new industries and to consumers through the Chamber of Commerce, the regional division of the New York State Department of Economic Development, housed within the Western New York Economic Development Corporation (WNYEDC);
(2) provide long-term strategic planning for the region's economy through the Greater Buffalo Development Corporation and the WNYEDC;
(3) coordinate the myriad of distinct economic development programs in the Industrial Development Agencies (Erie and Amherst), the city and county economic development departments, the Horizon Waterfront Commission, the Niagara Frontier Transportation Authority (NFTA) and the Buffalo Place downtown development corporation;

(4) stimulate and support research and development through the Roswell Park Medical Center and the State University of New York; and
(5) enhance the region's cultural facilities and quality of life through the Buffalo Philharmonic and the Greater Buffalo Development Foundation (GBDF).

THE NEW AGENTS OF REGENERATION: "HIRED GUNS" AND PRODUCT DIVISION SPECIALISTS

As discussed previously, the Buffalo 18 pride themselves on their ability to get the best managers for their respective firms. From the beginning they set out to make sure that the best managers for the "product divisions" of urban regeneration were hired for Buffalo. As one member of the 18 put it, "We will pay whatever it takes to get the type person who can manage the process of development" (Lipsey, 1989).

The group members have demonstrated that they can do what Stan Lipsey says. In the past five years they have been very active in hiring new agency heads for the key agencies, or "product divisions" of urban regeneration. They have organized and paid for the recruitment processes for almost all these agencies. They even helped in the recruitment of key personnel for the new County Executive.

The vast majority of these new officials are from outside the region: The head of research for the university was recruited from the Rand Corporation; the head of the medical center from New Mexico and, prior to that, the Mayo clinic; the director of transportation from the Toronto Metro System; and the new head of the state's economic development authority was a national economic development officer for the Department of Housing and Urban Development (HUD) in Washington. The new head of the Chamber of Commerce was recruited from New England and the president and CEO of the GBDF came from the Houston Economic Development Council.

THE HIRED GUNS

"I'm a hired gun, brought in to create change, through the change process" said GBDF president Andrew Rudnick, when asked about the role of himself and the other professionals brought to Buffalo in the past few years. If Rudnick sounded a bit like an ex-Houstonite in his direct assessment of the role of the new economic development specialist, Canadian Alf Savage, president of the region's transportation authority was just as direct:

> They have paid good money to bring us to town. We are here to create change. We are the new technocrats, who have been brought in to make policy because the political councils no longer make policy. In the past few years the city was run by the politicians and the result was a gray funk. With the appointment of people like Rudnick, Tobe (at the County) and myself, the leadership has voted for an aggressive, extroverted, management style which believes in rational planning and rational decision making. The pendulum has swung away from a city drowning in traditional politics to the era of the technocrats.

Another of these technocrats is the state's chief economic development officer in the region, Judith Kossy, president of the WNYEDC. Kossy, hired away from the federal government in Washington, agrees with Rudnick and Savage about their role in the regeneration process. She suggests that the activities of the 18 are more ones of corporate oversight and efficiency and less those of vision and management:

> My sense is that the 18 did not know what it meant to turn the region around in this time of crisis, except to influence the turn around through the institutions of economic development including the Chamber of Commerce, the GBDF, the Erie County Industrial Development Authority (ECIDA) and the WNYEDC. With no agenda, no staff and no formal constitution, they needed technical support and other ways to help form the agenda to focus investment and the priorities of the region. Our agencies are the forum for this and we are hired as the technical managers of the process [Kossy, 1989].

Every Monday afternoon at 3:00 p.m., these new agents of regeneration meet at the county industrial development agency office. Besides Andrew Rudnick, Judith Kossy, and Alfred Savage, the heads of the county's two largest industrial development authorities, Richard Swist and James Allen, university vice-president for Research and Sponsored Programs Michael Landi, Chamber president Kevin Keeley, county development director Richard Tobe, and city development corporation head Charles Rosenow are almost always present. There is rarely, if ever, a formal agenda. But what does transpire at these sessions has become an important part of the economic development process of the region. With the exception of their host, ECIDA president Richard Swist, Amherst Industrial Development Agency (IDA) head James Allen, and city representative Rosenow, all those attending are from outside the region. In the past, their agencies had been competitive and rarely communicated, except when major conflicts occurred or when they were publically thrown together on projects. The primary purpose for these meetings, according to Rick Swist of the ECIDA, is:

> To communicate—to get to know each other and diffuse the tension which naturally grows between us during the week. This meeting is something we at the IDA had proposed many years ago in our first master plan. Today, with these new professionals and with the commitment of the private sector, the meetings work [Swist, 1989].

The group or "gang," as they call themselves, uses the meeting to guard against what they all call "the problem of 'surprises.' We try to develop an understanding of one another's positions on issues even if we can't always develop a united front," says county development head Richard Tobe (1989).

The Monday afternoon "gang meetings" have become increasingly important in another way. If Kossy is right and the 18 don't have a specific action plan beyond hiring the best people to develop that agenda in the interests of local capital, then the gang meetings become the place where the region's regeneration agenda is hammered out. This is becoming increasingly clear to members of the gang. They have written and published the first "Greater Buffalo Economic Development Strategy (1987)" and linked this to each of the projects they are working on in their respective agencies. The Strategy is now a working document, reworked every year as the priorities of regeneration change. In the vacuum of political leadership in this area of economic development, this highly technical group of "outsiders" (or carpetbaggers, as Swist calls his peers) has developed the agenda of economic development for the region, wrapped it in the fabric of strategic planning, and "used the 18 to legitimize their plan" (Kossy, 1989).

THE "PRODUCT DIVISION SPECIALISTS"

The regeneration process also includes new as well as revised organizations which, while not directly concerned with economic development, have been deemed significant players in the urban regeneration game in Buffalo. Among these agencies are the Philharmonic, the city school system, and the Roswell Park medical center. The heads of these agencies are not "gang" members, not professional "hired guns" of economic development, but the Buffalo 18 has taken direct responsibility for each of these agencies. At present it has deployed Andrew Rudnick of the GBDF to be the acting director of the Philharmonic, and one of its members, Wilford Larson, president of Westwood Pharmeceuticals, to be chairman of the board. The 18 also sponsored a plan that joined all the medical centers in the region into a formal consortium and created the selection committee that hired the new director of the Rosewell Park research center, Dr. Thomas Tomasi. Most recently, the 18 agreed to underwrite the selection process for a new superintendent of Buffalo public schools.

WHITHER DEMOCRARY? THE RESTRUCTURING OF POLITICAL INFLUENCE

Every interviewee (politician, business leader, journalist, or administrator) agreed that elected representatives are becoming far less powerful in the process. If there has been an increase in the power of constituency, it has been in the executive branch: The two political leaders deemed essential to regeneration are the mayor and the governor. These two come as close as any political figures to having the power of a CEO. They have staff and agencies that can make far more important decisions than legislators at any level of government.

The essential weakness of legislators was described best by the economic development officer of Buffalo's sister city of Cleveland. He said "economic development policy must be done behind closed doors. You don't do deals by committee. Urban economic policy is by nature undemocratic" (Shumate, 1989). City and county legislatures have been given essentially peripheral roles in the economic development process—often reduced to arguing among themselves in an ongoing round of NIMBY actions. As such, the city council of Buffalo has been viewed as obstructionist because it represents neighborhoods with decidedly distinct race and class differences. They are representing increasingly old, Black, and Hispanic groups that have been neither the target nor the beneficiary of the economic development activities of the 18 or their "gang."

One gang member says that, for all this, there is still a place for county, and city legislators—at the end. These new agents of economic change make sure the legislator arrives in time "to cut the ribbon at the opening of each publically supported project [of regeneration]. After all, the legislative process still includes the appropriation function. We would be fools not to give them credit after the deal is done."

The implication here is clear. The recasting of urban leadership has led to a devaluation of urban democracy. Legislative participation in economic development is window dressing, or ribbon cutting. In Buffalo the vote really makes a difference only when electing a mayor with the political clout and the institutional independence to effect change.

All the organization, strategy, and new appointments in Buffalo have not been enough to overcome the continued economic fragility of the region. Economic regeneration in Buffalo has not produced a new competitive alternative to the traditional manufacturing base. In fact, the manufacturing base, while directly accounting for less than one-fifth of all the region's jobs, is still at the center of the region's long-term future. What makes this even more important is the fact that the manufacturing base of the region is precipitously skewed toward the auto industry. Put another way, if the auto industry were to leave the region, 105,000 jobs would be directly or indirectly threatened. This is about one-quarter of all manufacturing and service jobs in the Buffalo area (CRS, 1989).

The combined threats of a new recession, predicted by State Economic Development "czar" Vincent Tese (1990), and an almost certain retrenchment and further absolute decline of the U.S. auto industry (Cole et al., 1987; CRS, 1990) portend a serious new round of economic challenges for the region. There is little evidence that the "Buffalo 18," for all their organization, will be able to spare the region from economic decline.

REFERENCES

Anzalone, C. (1989). Reporter, *Buffalo News*. Interview.

Beauregard, R. (1989). Space, time, and economic restructuring. In R. Beauregard (Ed.). *Urban affairs annual reviews: Vol. 34. Economic restructuring and political response*. Newbury Park, CA: Sage.

Bingham, R., & Eberts, R. (1990). *The economic restructuring of the American Midwest*. New York: Kluwer.

Brown, R. C., & Watson, B. (1981). *Buffalo: Lake city and Niagara land*. New York: Windsor Publications, Inc.

Buffalo News. (1990, February 17). Editorial.

Campbell, D. (1989). Chairman and CEO, Computer Task Group. Interview.

Cole, D. E. et al. (1987). *The automobile endustry, General Motors, and Genesee County*. Unpublished report.

CRS. (1988). *The economic importance of the auto industry to Western New York*. (Working papers in Planning and Design). Buffalo: SUNY.

Eberle, S. (1989). Historian, Erie County Historical Society. Interview.

Feagin, J. R., & Smith, M. P. (1987). Cities and the new international division of labor: An overview. In M. P. Smith & J. R. Feagin (Eds.). *The capitalist city*. Oxford: Basil Blackwell.

Fuller, D. W. (1989). CEO, American Steamship. Interview.

Goldman, M. (1983). *High hopes: The rise and fall of Buffalo, New York*. Albany: SUNY Press.

Goldman, M. (1989). Historian. Interview.

Judd, D., & Parkinson, M. (1988). Urban revitalization in America and the U.K.—the politics of uneven development. In M. Parkinson, B. Foley, & D. Judd (Eds.), *Regenerating the cities: The U.K. and the U.S. experience*. Manchester: Manchester University Press.

Kensie, R. (1989). Former CEO, Goldome Bank. Interview.

Kossy, J. (1989). President, WNYEDC. Interview.

Kraushaar, R., & Perry, D. (1990). Buffalo: Region of no illusions. In R. Bingham & R. Eberts (Eds.), *The economic restructuring of the American Midwest*. New York: Kluwer.

Lipsey, S. (1989). Publisher, *Buffalo News*. Interview.

Magnate, M. (1989, March 27). How business bosses saved a sick city. *Fortune*.

Markusen, A., Noponen, H., & Driessen, K. (1989). *International trade, productivity, and regional job growth: A shift share interpretation*. Unpublished paper.

Perry, D. C. (1987). The politics of dependency in deindustrializing America: The case of Buffalo, New York. In M. P. Smith & J. R. Feagin (Eds.), *The capitalist city*. Oxford: Basil Blackwell.

Rudnick, A. (1989). President, GBDF. Interview.

Savage, A. (1989). Executive Director, NFTA. Interview.

Shumate, R. (1989). Office of Economic Development, City of Cleveland. Conversation.

Smith, M. P. (1989). *City, state and market*. Oxford: Basil Blackwell.

Sullivan, M. (1986, October 12). Group of 18 Strive to Boost Economy. *Buffalo News*.

Sullivan, M. (1989). Editor, *Buffalo News*. Interview.

Swist, R. (1989). Director, Erie County IDA. Interview.

Tese, V. (1990, January 30). Speech delivered to New York State Economic Development Zone Conference.

Tobe, R. (1989). Commissioner of Community Development and Environment, Erie County. Interview.

Voinovich, G. (1988). Keynote address delivered to the National Council of Urban Economic Development.

Wall Street Journal, The. (1988).

14

Regeneration in Marseilles: The Search for Political Stability

ANDRÉ DONZEL

FROM GROWTH TO CRISIS: ECONOMIC AND SOCIAL CHANGE IN MARSEILLES SINCE 1945

Marseilles, France's first port and second city, has traditionally been extremely sensitive to national and international economic changes. Throughout its history, Marseilles has exaggerated national and international fluctuations in times of both economic growth and crisis. It is hardly surprising, therefore, that the city so often proves to be an exceptional case in the French context.

1950-1974: THE GLORIOUS YEARS

In the decades after the Second World War, particularly between the mid-1950s and the early 1970s, Marseilles experienced dynamic population growth, which far exceeded the national trend and was without parallel in the city's history. Between 1954 and 1975 the city's population grew from 660,000 to 910,000, an increase of 38%, well above the national figure of 23%.

The background to this rapid change was the economic revival of the city and its region, illustrated by the increased volume of shipping through the port, the traditional barometer of local economic activity. Between 1950 and 1975, shipping multiplied by 10—from 10 to 100 million tons. At the same time, Marseilles consolidated its position as the largest port in France as its share of national shipping traffic grew from 20% to 33%. On a European scale Marseilles experienced similar progress. During the 1960s it moved from 5th to 2nd place in the league of European ports, behind Rotterdam, but ahead of Antwerp, Hamburg, and London. The rise was primarily explained by the role the port played in the international oil trade, particularly in processing and redistributing crude oil destined for West Germany, Switzerland, Italy, and Spain.

Marseilles' particularly favorable economic performance at this time was all the more unexpected since it coincided with Algerian independence, marking the end of French colonialism that had been the economic basis of the city during the nineteenth century. Hence, Marseilles' success cannot be understood without understanding the decisive role the state sector played in urban planning and industrial policy.

During the 1960s, as part of the Gaullist "equilibrium metropoles" policy, which attempted to create a framework of regional capitals oriented toward Europe, Marseilles and its surrounding region received massive industrial and urban development programs. The major achievement of this policy was the creation of a huge industrial and port complex at Fos-sur-Mer, about 50 kilometers outside the city. This was intended to provide the region with the heavy industrial base that it lacked and equip it with all the infrastructure required to make Marseilles "the Europort of the South" (Cultiaux, 1975; Tuppen, 1984).

While major national and international chemical and iron and steel companies were encouraged to set up on the new site, the center of the conurbation, where the region's main industrial activities had previously been concentrated, was redefined as the location for service-sector activities. Special emphasis was devoted to top management positions that the establishment of new companies in the region was supposed to attract.

A huge program for the renovation of the city center was developed with the aim of increasing office space and turning it into a business center. At the same time, a major urban redevelopment program was undertaken. Throughout the community territory, substantial resources were invested in housing, transport, an underground railway, expressways, commercial premises, hospitals, schools, and sports and cultural amenities to cater to a population that optimistic forecasts believed would reach 1,200,000 by 1985 (AGAM, 1975). Financed primarily by local public authorities, the policy led to soaring municipal expenditures by Marseilles for urban development, which contrasted sharply with the chronic lack of facilities that had characterized the interwar period.

1975-1989: PERIOD OF UPHEAVAL

However, the deterioration of the international economic situation after the oil crisis in 1973, and the shift of national and local policies toward fiscal constraints, meant that the driving force behind Marseilles' growth after the

Second World War was suddenly called into question. In the space of a few years, all the symptoms of a serious crisis accumulated in the city, which resulted in the present political turbulence of a kind that it has hardly experienced since 1945. The first sign of a reversal of the postwar trends was the start of a demographic decline, all the more remarkable since, until then, it had been an unknown phenomenon in the city. Traditionally a city of immigration, formed by successive waves of migration from all parts of France and the Mediterranean, since the mid-1970s Marseilles has experienced a decrease in immigration and a distinct drop in its resident population. Between 1975 and 1989, Marseilles' population fell at an accelerating rate. It lost around 5,000 inhabitants every year between 1975 and 1982; that figure has since reached 8,500 a year.

Of course, this phenomenon is not peculiar to Marseilles; all major urban centers are experiencing a decrease in population. But Marseilles covers a surface area two-and-a-half-times greater than Paris, and since it is only half-urbanized, the decline is clearly not attributable to the saturation of the area available for urban growth. Instead, it is the direct consequence of the decline in employment in the city in the past few years, particularly in the industrial sector. Having remained stable up to 1962, Marseilles lost 36,000 of its 100,000 production and construction industry jobs between 1962 and 1982. This hemorrhage has again worsened throughout the period: 2,000 jobs were lost between 1962 and 1968, 9,000 were lost between 1968 and 1975, and 25,000 jobs were lost between 1975 and 1982.

At the same time that industrial employment was declining, service-sector employment, which until then had compensated for decline, in turn succumbed. Having created more than 50,000 positions in the sector between 1962 and 1974, the city lost 10,000 jobs between 1975 and 1982. This decrease in employment was particularly severe in the city's port-related industries. But for the first time, it also affected jobs in the public sector—administration, welfare, health services, and education—which until then had witnessed enormous growth.

One obvious consequence has been a rapid rise in unemployment in the city: from 3.5% in 1962, to 14.2% in 1982, and 16% in 1987. However, these are average figures; much higher rates are found in the traditional industrial areas of the city. The result is a widening gap between these areas and the rest of the city. In terms of income, measured by the annual household tax declarations, the existing large disparities between the traditionally working-class, northern areas of the city and the traditionally middle-class, southern areas have grown. The three municipal districts in the southern part of the town, which represented 37% of the total income of Marseilles' households in 1977, represented 42% by 1986. Conversely, for the three northern municipal districts, despite a comparable population level, the figures were 9% in 1977 and 7% in 1986.

There is little prospect of the position's improving in the short term since the pressures on local public finances make it less possible to reduce these disparities. The loan repayments contracted during the growth period now weigh heavily on the city council's budget, while its revenue expenditure is very difficult to reduce. The result is that for several years, there has been a marked slowing down of expenditure on major urban development programs. This is particularly clear in housing: After reaching 10,000 dwellings per year in the 1960s, the rate of construction in Marseilles dropped to less than 2,500 dwellings per year after 1974. For other types of amenities, with the exception of the underground railway—which has to be completed to avoid increasing its management deficit—there is a similar trend. After 1977 the city council announced its intention to suspend its development policy by replacing the "policy of major works" with a "major policy of small works."

In view of the present trend to cut back government subsidies to local authorities, and the reduction of the fiscal resources of the city council as a result of the demographic depression, public money cannot continue as the main stimulus for growth. At the same time, the economic crisis has had a long-lasting effect on Marseilles' economic potential and has compromised the ability of the local private sector to take the place of public initiative. Under these conditions, there is no other solution than to depend on external decision-makers if the city wishes to recover economically. All these obstacles and uncertainties were bound to have major repercussions on Marseilles' internal political equilibrium.

FROM DEFFERRE TO VIGOUROUX: A TROUBLED POLITICAL LANDSCAPE

Ten years ago, Marseilles entered a phase of political instability that it had hardly experienced since the Second World War. After 1945, despite certain internal shifts on the left and the right, the city had been very stable politically, apparently unaffected by either the socio-demographic changes that the city was undergoing or the fluctuations in national political circumstances. Without fail at every election, Marseilles repeated its internal geopolitical pattern, falling into three large political strongholds—communist in the North, socialist in the East, and right-wing in the South. This distribution of votes reflected the segregated social structure of Marseilles, with the working-class and upper middle-class areas at opposite ends of the city, and the middle class between the two. Out of this precarious balance a political personality had emerged, who, by constructing a series of changing alliances, was able to dominate the local political scene for more than 40 years. This was the mayor of Marseilles, Gaston Defferre.

THE DEFFERRE YEARS

Defferre originally headed the municipal government immediately after the Liberation, from 1944 to 1947, with the support of the communists. He broke this alliance to stand in the council elections of 1953, this time with the support of a part of the Marseilles bourgeoisie. Despite the fact that he belonged to the Socialist Party, they saw him as an effective "buffer against communism," which was then very powerful in the city. On this basis, a power system took shape that was quite unique in view of Marseilles' political tradition and the national postwar political context in France.

A Comeback for Local Autonomy

Throughout its history, Marseilles has experienced great difficulty in finding a real political elite capable of heading the city in an efficient and durable manner. Although possessing all the attributes of an independent state when founded by the Greeks in the 6th century BC, the whole history of this city is a continuing process of political beheading, resulting in most cases in its being divested of its administrative powers. The Ancien Regime, which was suspicious of a city that had maintained its republican traditions from antiquity, even went to the extent of defending the city by pointing the guns, not toward the outside to deter would-be aggressors, but to the inside to subdue a population that was little inclined to depend on the Kingdom of France.

In time, the Republic would not treat the city with much more regard. Marseilles paid very dearly for its federalist sympathies and its opposition to Parisian centralism during the Revolution. A special administrative regime was imposed, which deprived it of all elective institutions and placed it under the authority of prefects nominated by the central power.

This tutelage regime, which did not really come to an end until after the last war, apart from certain periods when it was suspended, hardly encouraged the generation of political leadership with a well-organized local base. This was all the more true since the city's economic elites for a long time showed little interest in participating directly in local political institutions. Marseilles' capitalist structure, with its predominance of a bourgeoisie that was more commercially than industrially oriented, and thus attached to economic liberalism, did not encourage such participation. Moreover, the presence of a large work force, which was constantly renewed by massive waves of immigration and perfectly suited to casual unskilled work, did not, in the eyes of the local employers, justify paternalistic politicians taking responsibility for the local workers.

These conditions explain why, for a long time, social groups grew up around corporate organizations representing their particular interests, rather than around municipal institutions. On the employers' side, this was particularly demonstrated by the powerful Chamber of Commerce, which for a long time eclipsed the local government in the management of city affairs, because of its role in the management of the port. But the workers were also well represented by all the organizations capable of catering to their interests: unions, mutual benefit societies, local committees (Sewell, 1984). Equally, their early interest in "Guesdism," the first example of the communist current in the French workers movement, did not incline them to turn to local structures, since national struggles had long taken precedence over local political action (Bleitrach & Lojkine, 1981).

This resulted, however, in the municipal structure's remaining underdeveloped with a pitiful reputation until the eve of the Second World War. Anarchy, incompetence, and corruption were the most common features of the prewar Marseilles municipal scene. This was especially true in the 1930s, when a Mussolini-type figure with underworld connections made use of questionable administrative methods, earning Marseilles the name "French Chicago."

In view of the city's past, the Defferre legitimacy stemmed from his ability to eliminate its reputation as an ungovernable city by imposing more orthodox administrative methods, more in line with the standards dictated by the Parisian Ministries. From 1953, the municipal budgets were regularly balanced. In 1959, even before national legislation made it compulsory for all French communities, the city implemented a plan regulating its development. Progressively, the city assembled a whole range of technical, administrative, and financial facilities that gave it control over development—an urban development agency, mixed investment companies, and a municipal computer center.

Defferre's Third Way

But this new ethic in the running of municipal affairs would not in itself have been enough to ensure the longevity of the new power structure if it had not been accompanied by a major change in Marseilles' social structure. The decline of its traditional port role resulted in the polarization of the local bourgeoisie and the eclipsing of the working class. After 1954 there was a continuing decline in the proportion of both workers and unsalaried workers in the working population. Conversely, there was a considerable increase in the intermediate categories, in particular middle and top managers whose number doubled between 1954 and 1975, and whose share of the economically active population increased from 15% to 27% in the same period.

These new rapidly rising social strata progressively assumed an important function in the development of the city, asserting themselves as the new leaders with the future of urban Marseilles in their hands. This transition occurred with a considerable expansion of the role played by the municipality. Whereas until the 1950s the real power behind the development of Marseilles was the Chamber of Commerce, the second half of the twentieth century saw municipal institutions emerge as the strategic ground for decisions about urban transformation and the point at which compromises between the various social forces were made.

This transformation was accompanied on the electoral front by a growing predominance in local politics of the Socialist Party, led by the mayor. By 1977 the Socialists held a majority of seats on the city council, without the support of outside political parties which had until then been necessary. This was the peak of the Defferre System, which, incarnating a "third way" between Communism and Gaullism, seemed to be an alternative to the political bipolarization that had been a feature of French society since the last war. But it was at the very moment it seemed so firmly established locally and nationally when, after Mitterrand's election as President in 1981, Defferre was made a government minister, that the system began to show signs of weakness.

FROM GASTON DEFFERRE TO ROBERT VIGOUROUX

Although a variety of indicators pointed to the fact that Marseilles had entered a phase of chronic depression in the middle of the 1970s, the mayor, who until then had been seen as the principal architect of its renewal, began to appear in the 1980s as the perpetrator of its decline. Converging criticism from a wide sector of opinion seriously threatened his political base, forcing him to accept an alliance, which he had long rejected, with the Communist party in the city council elections in 1983. Defferre gained a narrow victory over his right-wing challenger, Jean-Claude Gaudin, who had formerly been part of the municipal majority.

The Right blamed him for not having reacted strongly enough to the exodus of companies from the area and for not having created a favorable climate to attract the new high-tech industries. The Left denounced his inegalitarian urban policy, which, by concentrating public housing in the north of the city, simply reinforced the existing social segregation between the northern and southern areas of the city, creating veritable ghettos in its midst. The Socialist Party itself also began to contest the pyramid-like, authoritarian organization of power in the local party, which had prevented the emergence of new political leaders, in marked contrast to developments on the national level.

After 1981, however, and especially after 1983, the mayor, at the head of a new majority that united Socialists and Communists for the first time since the Liberation, took heed of his critics. He implemented a series of measures aimed at curbing the deindustrialization that had been taking place in the city for decades. An economic mission was created to promote the city in the eyes of industrialists and encourage their establishment in Marseilles by granting them land and financial incentives. In the Chateau Gombert area in the northeast of the city, a technological center consisting of 180 hectares was constructed to provide Marseilles with a facility that would attract the new technological industries. At the same time, a vast public housing rehabilitation program was undertaken in the outlying areas, aimed at restoring a fabric that was becoming more and more dilapidated. Efforts were also made to improve the living conditions of the population by creating amenities that were often lacking—cultural centers, parks and gardens, new roads.

There was a particularly striking example of self-criticism on the part of a mayor who enjoyed the reputation of being one of the greatest centralizers in France. As Minister of the Interior, Gaston Defferre became the originator of a major decentralization law that aimed at reinjecting life into local democracy by giving extended powers to city councils. This law in particular obliged the large French cities—Paris, Lyons, and Marseilles—to set up structures to bring them closer to the inhabitants—municipal district councils, whose elected representatives have the right to express their opinion on all decisions taken by the central city councils about their district.

These new measures were not, however, enough to check the electoral erosion of the new municipal majority created by the 1983 elections. Subsequent election confirmed the waning influence of the traditional left-wing parties in favor of the traditional Right. Increasingly, an extreme right-wing movement that had recently arrived on the local and national political scene, the National Front, exploited the decline of the Left.

The Le Pen Phenomenon

The strong support for the National Front in Marseilles—28% of votes on average since it appeared in the European elections of 1984—once more singled out the city in the national political context, particularly since the leader of this party, Jean-Marie Le Pen, chose Marseilles as his main battlefield. The rise of the extreme right in Marseilles, a city that was traditionally left-wing, left many observers puzzled. Indeed, feeding mainly on the rejection of immigration, it appears an anachronistic reaction to a phenomenon that stabilized a long time ago in Marseilles. Immigration has even slowed down in the past few years, and the city now has the lowest proportion of foreigners that it has ever known in its history (Donzel, 1989). Contrary to the impression given by its extreme ideology, the National Front found its key support in the lower middle-class "with social mobility problems, who have begun to rise in the social scale but whose status still remains uncertain, the 'small whites' who feel threatened with regard to employment and security . . . and who are therefore more readily anxious and intolerant" (Bon & Cheylan, 1989). While it demonstrates the disarray of part of the Marseilles population, the "Le Pen phenomenon" has not been the only destabilizing factor in local political life in recent years.

The War of Succession

The sudden death of Gaston Defferre in May 1986—following a meeting of the local Socialist Party, during which he was placed in a minority by his chief lieutenant—provoked a war of succession in Marseilles' corridors of power. The affair resembled a Greek tragedy that held the whole of France in suspense for several months. Choosing a successor to a political personality such as Gaston Defferre, who had made his personal mark on local and national politics for more than 40 years, was bound to be a painful process. This was especially true since the power system he had constructed throughout his long career went well beyond the normal municipal institutions and covered a massive complex of organizations, extending from the Socialist Party, with its numerous outposts in public life and the local community associations, trade unions, to the regional press, whose main headlines were under his control.

At first it appeared that the struggle for succession would take place among those in the corridors of power, who seemed capable of continuing the "Defferrist" policy in his three strongholds: the Town Hall, the Socialist Party Federation, and the "Le Provencal" press group. It particularly pitted against one another those who had established their political credentials either within the municipal machine as part of their functions at the head of the city's administration, or within the Socialist Party as part of their responsibilities as political activists. Hence, Philippe San Marco, the general secretary of the mayor of Marseilles, and Michel Pezet, secretary of the Socialist Federation, were for a long time the potential heirs to the mayor. The late mayor's wife, Edmonde Charles-Roux, a renowned novelist and editor-in-chief of the "Provencal" who held trump cards that could decide the outcome of this transitional phase, was also a contender.

But what at first seemed to be an internal power struggle of local interest soon extended beyond the bounds of Marseilles. In national politics, people have not forgotten the proverbial saying of Louis XIV, "He who holds Marseilles holds France." This was even more true for a socialist government because Marseilles was the only large French city led by a socialist majority. There was a general feeling that, in a city with a strong republican tradition, the dynastic quarrels might alienate the electorate and accelerate the party's worrying electoral decline. In the days after the mayor's death, attempts were made by the national leadership of the Socialist Party to encourage a truce between the would-be successors. The succession was to go to a personality who was sufficiently close to the lay mayor to guarantee the continuation of Defferrism—and sufficiently inexperienced to avoid compromising the most likely heirs' chances of being in the right strategic position in the battle for the conquest of Marseilles, for which the decisive point was the next council elections in March 1989.

The ideal "regent"—an "interim pope," as he was then christened—responsible for administering the city until that date, was found in the person of Robert-Paul Vigouroux. A personality little-known to the general public, except perhaps for his work as a neurosurgeon in the city's largest hospital, he had until then only occupied minor posts in the city cCouncil or in the Socialist Party. He hardly seemed capable of having a decisive influence on the local political arena and did not appear to have enough political weight to preside over the destiny of France's second city for a long time.

The Vigouroux Stamp

However, two factors helped to radically modify this impression. What many never suspected at the outset was the considerable personal impact of this new mayor on the Marseilles population. A good doctor, a good husband, a good father, he closely resembled a former figure in prewar municipal life, who incarnated a tradition of populism peculiar to Marseilles and who became a veritable cult figure in the city—Mayor Siméon Flaissiées. He also was a doctor and an outsider, who despite the vicissitudes of the era had succeeded in creating the "sacred union" of the Marseilles people (Donzel & Hayot, 1989). This growing popularity would, however, have never been enough to ensure the preeminence of the new mayor in local public affairs, without strong backing from local and national political circles.

At the national level, a number of people, with the President of the Republic at their head, seized the opportunity to exploit the "Vigouroux effect," not only for immediate electoral results but also as a long-term political strategy. Following the rupture of the Union of the Left at the national level after the exclusion of the Communist Ministers from the Government in 1984, Mitterrand was looking for a new political angle to enable him to increase his majority on the eve of the political elections of

1988. This new strategy took shape as *ouverture à la société civile ("include society at large"). In a political climate more favorable to liberal policies than in 1981, it aimed at fostering close collaboration between the leaders of the private sector in public affairs.*

Given the political recomposition that was taking place in Marseilles, the city could become the testing ground of this opening up to "society at large." Vigouroux would have little difficulty in adopting this strategy as his own, in view of the fact that he had no support from his own party on the local level. Indeed, he had no alternative but to endorse it. In fact, this system had, in some ways, already been implemented with the arrival in this city in 1985 of a personality who was to become one of the key figures of this new presidential strategy—the business tycoon and future member of parliament, Bernard Tapie.

The son of a worker, born in the suburbs of Paris, he acquired short-lived glory in his youth as a pop singer. But his talents were later revealed in quite a different field. He was to soar to fame in the business world by buying up struggling or bankrupt companies at a low price, which he reorganized and turned into viable ventures. Riding the wave of success in the business world, he became the head of a huge conglomerate of companies from industry to the media and he soon became, in the 1980s, the symbol of the "winner" in France.

This expert in lost causes could not help but be interested in Marseilles. More than anyone else, he realized the potential offered by this city in various fields. He made his entry in a rather unexpected manner, by buying up one of the most prestigious and profitable enterprises of the city—its football club, "l'Olympique de Marseille." Since 1981, this football club, the oldest and most titled in France, had been going downhill in the second division, to the great despair of the Marseilles population. Within three years, with the arrival of new players and management recruited by Tapie at very high cost, the OM not only climbed back into the first division, but won the French Championship.

Following this spectacular salvage job, Tapie enjoyed immediate political recognition, which was all the more legitimate because the supporters of the club are known as the First Party of Marseilles. In spite of the reluctance of his local allies, Francois Mitterrand was all the more willing to support him because this representative of "free enterprise" had enough clout on the national level to sanction, in the business and economic community, the president's wish to open up to society at large.

The advent of Tapie on the Marseilles political scene also favoured Vigouroux. The ease with which Tapie was elected a member of parliament in Marseilles in 1988, despite the opposition of most of the local political machine, demonstrated the volatile nature of the electorate. From this moment, Robert Vigouroux, whose candidacy for the mayoralty many considered doomed to failure, was to gain credibility. The forces that until then had been dormant in the wake of this virtual loser, were to break through to provide him with support far beyond that mobilized by Defferre. Progressively, the embryonic "Vigouroux Network" was to have ramifications far and wide, locally and nationally. On the evening of March, 19, 1989, Robert Vigouroux was triumphantly elected the head of the city council of Marseilles. All the other candidates, both on the Right and the Left, suffered heavy losses. Apart from the National Front, the victim who suffered the most spectacular setback was Michel Pezet, the official candidate of the Socialist Party. Even now he is haunted by the words of Gaston Defferre at the Socialist Party congress: "Well, if no one else will say it, I will—Mitterrand is the grave-digger of socialism in France" (Raffy, 1989).

THE STAKES OF URBAN REGENERATION

It is still too early to assess the effects of the changes in the city's politics since Vigouroux was not very explicit in his campaign about his plans for the city. Nevertheless, three major concerns, which are shared by many local politicians, have emerged.

The first is the role of the port and manufacturing activities in the city's future as it faces an ever-deteriorating employment problems. The second relates to the increasing disparities in the conurbation between the center and the periphery, old and new areas. The third is the method by which to control these developments—the place of private and public investment, and the role of local and national actors, the choice of government or nongovernment as lead players in urban regeneration.

THE SEARCH FOR NEW GROWTH

"Marseilles is a port." That is a truism which nobody denied until very recently. The port was in the heart of the city; it was the very reason for its existence. But since the port facilities drifted northwards, it is now far away from its original site. For the most part, the port's activity is not concentrated in Marseilles, but in Fos-sur-Mer, which boasts the huge half-deserted spaces of la Crau. Marseilles basin itself only accounts for 6% of the total traffic of this new port complex. If petroleum products and bulk cargo are excluded, the significance of the Marseilles basin for general cargo has been challenged for several years. They contributed only 49% in 1987, compared to 63% in

1984. Industrial investment has undergone parallel development. To the west, the areas of Fos and the Etang de Berre now absorb most investment, following the location of large national and multinational firms in petrochemicals (Shell, BP, Total, Esso), iron and steel (Solmer, Ugine Aciers), and heavy chemicals (ICI, Pechiney, Atlantic Richfield).

At the other extreme of the port structure, Marseilles, the former center of gravity of the regional economy, has not only failed to benefit from the spill-over effect from the Fos area, but it is also being rivaled by the new enterprise zones set up in the outlying areas such as Aix, Vitrolles, or Aubagne. By offering the companies more attractive conditions, cheap land, easy access, and pleasant environment, they have contributed to the exodus of firms from the center of the conurbation. This vitality of the peripheral areas is all the more worrisome for Marseilles, since it involves not only its industrial activities but also those that were to inject life into its redevelopment—research, the service-sector, high-tech industries.

The 1980s were marked by a number of initiatives aimed at making the city more economically attractive, with some rehabilitation of its traditional port functions. Measures to regenerate industrial wastelands were taken as the city purchased them to maintain an acceptable quality of land within the city boundaries. Areas called "enterprise villages" were set up to accommodate small and medium-size businesses with conditions comparable to those offered by the peripheral areas. In partnership with the Chamber of Commerce and the Regional Council, the city embarked upon the creation of the "Chateau Gombert Technological Centre," destined to place Marseilles on the "high-tech route" linking the main high-tech centers of the southeast of France from Nice to Montpellier.

At the same time, serious thought was given to the future of the specifically Marseilles part of the port. Not so long ago, the technical and economic development of maritime traffic seemed to favor Fos, to the detriment of the Marseilles basins. The maintenance was even said to be an obstacle to the tertiary development policy undertaken by the city in the 1960s and 1970s, and there was even talk of remodeling them along the lines of the London Docklands. But the respectable performance in certain sectors of port-like general cargo traffic and the emergence of new off-shore industries and high-performance companies in Marseilles—and the essential role of these basins in the port structure—encouraged the need for caution.

In 1989 the port authorities reinforced their maritime role, as part of a master development plan for the basins, by the extension of its water area to accommodate the supercontainers. The Chamber of Commerce itself, in the past one of the most fervent promoters of Fos, is so convinced of the revival of the Marseilles basins that it has offered to add 1,000 hectares of extra land assigned to industrial activities (CCIM, 1989). This optimism is not as unrealistic as it seems, since former industrial sites of the Marseilles region are enjoying a new lease on life. This is especially the case of the La Ciotat Shipyards, which recently were to be converted into a beach resort, but are now to reopen on the initiative of the Swedish-American backed company, Lexmar.

RECREATING A COMPLEX URBAN FABRIC

The new recognition of the role of the industrial sector has reintroduced greater diversity in the city's approach to urban planning and development. Marseilles has long enjoyed the reputation of being a "federation of villages." Historically, the city was constructed on the basis of small urban communities—first parishes, then neighborhoods—which remained relatively close to the place of employment and place of residence. This naturally led to the formation of very close social relationships among the inhabitants. Until very recently, people from Marseilles did not claim to be "from Marseilles," but rather from St. Antoine, St. Pierre, la Cabucelle, Ste. Anne, or le Redon. At the same time, a highly developed road system served the vast community areas, heading directly toward a single center in the old port and the Canebiere, the traditional meeting point and melting pot of the city. On this basis, Marseilles, shaped by the influx of immigrant populations, produced a common identity of sorts.

However, the schemes of the functionalist architects of the 1930s, which had a major influence on postwar municipal urban policies in Marseilles (Donzel, 1983), challenged this pattern of urbanization. With the urban expansion of the 1950s and 1960s, the city, which until then had experienced little segregation between the place of work and social activities, developed a "zoning" system that clearly opposed employment and residential areas, with the latter graded into private residential areas in the south and rented accommodation in the north (Donzel & Garnier, 1989).

From the mid-1970s onward, efforts were made to counteract this fragmentation of Marseilles' urban fabric. The creation of an underground system, the revitalization of the city center, and the rehabilitation of the public housing stock in the peripheral areas were all implemented with this in mind. The city also took a multitude of cultural initiatives, making Marseilles one of the most attractive French towns in this field. However, this attempt to rebuild a more socially and functionally balanced urban environment remains in the grip of institutional and financial machinery that is ill-suited to the new policy.

RESOURCES FOR REGENERATION

The cost of urban regeneration means Marseilles has to rely upon external sources of income. Fully aware of this fact, some local politicians are making a concerted effort to make the satellite towns, which generally have more fiscal resources, play a greater role in the future of Marseilles. This has brought forward the idea of a "Greater Marseilles." Some have made this a "national cause" by asking the state for greater financial commitment to local development(San Marco & Morel, 1988).

Following the creation of an "Interministerial Delegation for City Affairs" in 1988, the Prime Minister, Michel Rocard, revealed his intention of making urban development one of the priorities of his government "to enable the cities to be in the Europe of tomorrow, the representative of social solidarity and the force behind development." This has been translated into a new type of partnership between the state and local government with "city contracts," in which the state undertakes to contribute to the financial needs of certain cities for several years. While more selective but more substantial than the former types of state aid, the policy requires the cities to define "global development projects," which require active partnership between the local protagonists—private companies, local administration, and various associations. In 1989, 13 French cities benefited from this procedure, including Marseilles. In the terms of this policy, Vigouroux has undertaken to complete, within six years, 50 projects that have top priority for the city's economic development—the modernization of the port in the particular—or its urban development, the extension of the underground system, motorways, continuation of the rehabilitation of the urban stock.

But this development does not in itself dispel the feeling that there is a marked trend for the central state to move away from financing local government. This explains the real willingness of the new mayor to find other partners for the city. With this in mind, he has been one of the initiators of the "Eurocity" network, which has assembled 21 large European cities to represent the interests of the large cities with the European Economic Community. This "European cities lobby" arose from the need to reestablish equity in the EEC's budget, which the cities believe places too much emphasis on rural areas while being financed by the large cities.

The last elections also created renewed interest, on the part of the business sector within the city, with the birth of "the Marseilles-Provence Foundation." This consists of the 20 main economic actors with links with Marseilles. It includes the representatives of large local firms as well as other outside decision-makers who wish to invest more in Marseilles.

A DEMARCATION LINE

After 15 years of economic crisis, Marseilles can now see the light at the end of the tunnel. But the crisis has left an indelible mark on the city. Unemployment and instability of employment still loom on the horizon for a large part of the population, especially young people. At the same time, the gap between the rich and poor areas is widening. The Canebiere, once the area that used to "incorporate" the city, has become, in residents' eyes, a demarcation line that increasingly divides the north and the south of the city. These social divisions make it less possible to imagine the renewal of a common urban consciousness in the short term. Moreover, the political changes experienced by the city have left deep wounds on the local political world. If the record of the political parties of this city can be criticized, it is not certain that the lobby groups that have taken their place are a better guarantee of political equality. The years to come will tell if Marseilles, whose long history has been one of fabulous phases of expansion as well as spectacular setbacks, is indeed on the way to recovery.

REFERENCES

Agence D'Urbanisme de L'Agglomeration Marseillaise. (1975). *Schéma directeur d'aménagement et d'urbanisme de l'agglomération marseillaise*. Marseille: AGAM.

Bleitrach, D., & Lojkine, J. (1981). *Classe ouvrière et social-démocratie; Lille et Marseille*. Paris: Ed. Sociales.

Bon, F., & Cheylan, J. P. (1988). *La France qui vote*. Paris: Hachette.

Chambre de Commerce et D'Industrie de Marseille. (1989, November). *Marseille Provence prospective*. Marseille: C.C.I.M.

Cultiaux, D. (1975, February). L'aménagement de la région Fos-Etang de Berre. *Note et etudes documentaires*. Paris: La Documentation Française.

Defferre, G. (1971). Marseille, le passé, l'avenir. In revue *Marseille*. Ville de Marseille.

Donzel, A. (1983). *Marseille, politique urbaine et société locale*. Thesis at the University of Provence, Aix en Provence.

Donzel, A. (1988). Comportements politiques et immigration; le cas de Marseille. In *Des migrants et des villes, mobilité et insertion*. Travaux et documents de I'IREMAM, n°6, Aix en Provence.

Donzel, A., & Garnier, J. C. (1988). *Des lieux à histoires: Marseille Quartiers Nord*. Marseille: CRES-CNRS.

Donzel, A., & Hayot, A. (1989, April, May, June). Marseille, l'effet Vigouroux. Tradition historique et recomposition politique. *Avis de Recherche, n°17*.

Raffy, S. (1989). *Les enfants de Gaston. La bataille pour Marseille*. Paris: Editions Jean-Claude Lattès.

San Marco, P., & Morel, B. (1988). *Marseille: l'Etat du futur*. Aix en Provence: Edisud.

Sewell, W. H. (1974, November). Social change and the rise of working-class politics in nineteenth-century Marseille. *Past and Present, 65*.

Tuppen, J. N. (1984). The port-industrial complex of Fos: a regional growth pole? In B. S. Hoyle & D. Hilling, (Eds.), *Seaport systems and spatial change*. John Wiley and Sons, Ltd.

Conclusion

15

Patterns of Leadership

DENNIS JUDD
MICHAEL PARKINSON

THE DEVELOPMENT OF LEADERSHIP

Leadership capacity is affected by the skill with which leaders exploit the resources available to them. Those resources include such things as the strength and durability of the public and private sectors and political relations between them; the availability of national resources like money, goodwill and legislation; the local social structure, that is, class relations, social solidarity, and entrepreneurial traditions. All of these characteristics help determine the complexity and sophistication of the strategies that are adopted to regenerate the local economy. More complex, proactive strategies support a diverse range of growth sectors, involve a large number of institutional, political, and community leaders, and rely on relatively inclusive coalitions. More "primitive" strategies, such as those adopted in Houston, try to create a playing field that investors will want to occupy, and rely on very general, reactive policies, such as tax incentives and low labor costs, and leave most important decisions and initiative to private actors.

Leadership is a crucial variable in determining how cities respond to economic change. Rennes, Pittsburgh, and Hamburg possess the greatest capacity to define and sustain complex strategies of economic revitalization. The institutional fabric, the traditions of leadership, and the political coalitions that sustain them are sufficiently well developed that, even when they make mistakes and experience failures, they recover, adapt, take a new direction, and press on. Thus, Pittsburgh has gone through three distinct phases in its redevelopment history. The elites in Rennes were able to deftly adapt from a very centralized political system, which required them to seek influence with national political elites, to a historic, radical decentralization that required them to define their own future. In the 1970s, elites in Hamburg miscalculated when they supported unprofitable industries, but in the 1980s they changed direction and focused their strategies on new sectors.

At the other end of the spectrum, Liverpool, Buffalo, and Marseilles failed to produce a leadership that could identify and pursue regeneration strategies even as ambitious as Houston's. Even under extreme pressure Marseilles was unable to overcome its political fragmentation; partly because power was traditionally centered in the hands of a leadership that exercised a machine-politics style of control over local decisions of all kinds, and partly because the city was divided socially and ethnically. As a consequence, the local political system did not develop the capacity to facilitate brokering among the city's ethnic, racial, and social groups. Leadership in Marseilles remains divided.

Our case studies have shown that leadership is not a static but a developmental phenomenon. Past experience matters a great deal. In cities with a history of development efforts, political battles tend to encourage the creation of a rich institutional context. The political struggles over development, which are more or less inevitable, in some cases led to the evolution of a more inclusive style of local politics, since if political leaders are sophisticated, they soon co-opt protesters into the decisional circles. Whether this means that the benefits of regeneration also get distributed more widely and equitably is an important question, to which we will return later.

LEADERSHIP CAPACITY AND REGENERATION STRATEGIES

The experiences of cities profiled in this book reveal that leadership determines how cities respond to economic decline. Although economic circumstances impose constraints, cities that have developed substantial leadership capacity have been able to develop more complex regeneration strategies than cities where leadership is weak and fragmented. As Figure 15.1 illustrates, there are connections between leadership capacity and the complexity of regeneration strategies.

The case studies show there are different routes to leadership capacity, with the public and private sectors playing different roles in different cities. However, both sectors are important. Where either sector is weak, or where there is conflict between them, the task of responding to change and shaping the future becomes more difficult. It is significant that these generalizations hold, despite major differences in party systems, ideological traditions, and constitutional and financial arrangements between the center and local government. In the U.S., Canada, the U.K., France, and West Germany, the

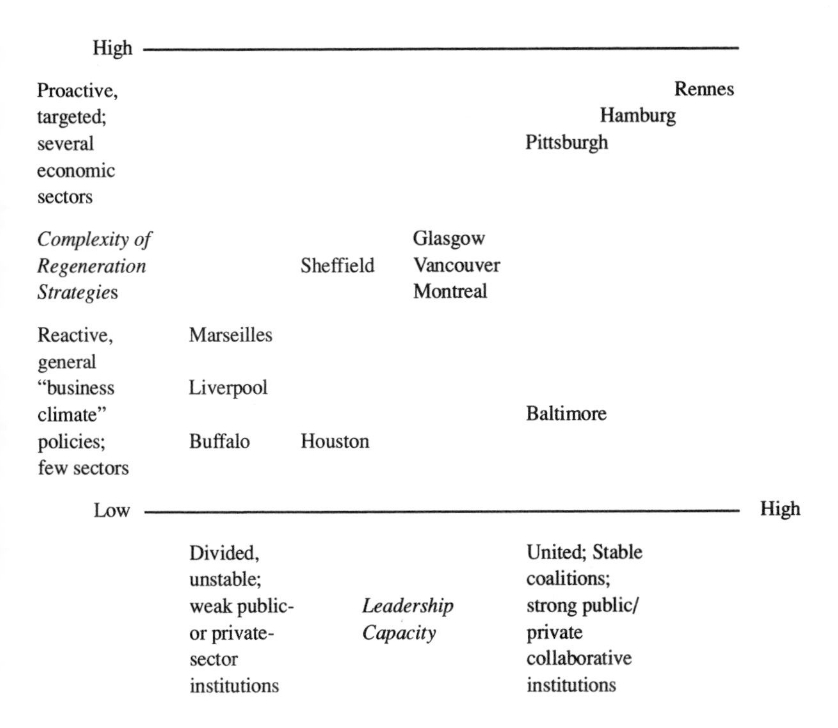

Figure 15.1 Leadership Capacity and Regeneration Strategies in Twelve Cities: A Map

leaders of cities began to reevaluate their economic regeneration strategies at about the same time, from 1980 to 1983. Perhaps most striking is that, in two very different systems on opposite sides of the Atlantic, Rennes and Pittsburgh evolved very strong leadership coalitions that revolved around aggressive public-sector elites, but also in these countries, there were cities, in this case, Buffalo and Marseilles, which did not successfully respond to economic crises.

Cities vary systematically in the choice of their regeneration strategies. The three most successful cities profiled in this book have diverse social and economic structures with substantial numbers of qualified people based around thriving sectors, which allowed them to coordinate complex regeneration strategies. In general, cities adopt less complex strategies either when they have relied for too long on declining sectors or when they have become dependent on a single sector. Houston's dependence on oil is one case, and Baltimore's reliance on tourism is another. More complex strategies that incorporate a range of high value-added sectors are more likely to lay a solid foundation for sustained regeneration. But leaders are very constrained in this respect by the resources available to them, and in particular by the social structure of their cities. In the short term, at least, cities like Liverpool and Buffalo do not have rich resources and, hence, they have fewer options. Even in the long term it is not clear that such cities have the powers or resources to make the systematic investment in human capital required to sustain a coordinated, complex regeneration strategy.

Pittsburgh, Rennes, and Hamburg have been most able to build a complex, institutionalized leadership structure committed to the revitalization of the local economy. As Alberta Sbragia points out, in Pittsburgh the several decades of experience with local economic development "has developed a public sector that is sufficiently powerful and professionalized to act as a true 'partner' with the private sector." The strategies and coalitions that support economic regeneration rest upon a foundation of many earlier efforts. This may make Pittsburgh exceptional in the U.S. context. It certainly does not fit the "private city" model that normally is used to describe public-private relationships in American cities. The formal incorporation of the nonprofit sector into Pittsburgh's economic development has been made possible through consistent and persistent public leadership. The inclusion of neighborhood leaders and organizations, as well as business leaders, in economic development planning also has been made possible by the ability of public officials to articulate an agenda and define a process of negotiation and coalition-building.

Pittsburgh demonstrates how much difference leadership can make to the process of economic regeneration. The almost complete deindustrialization that occurred in Pittsburgh and in its surrounding region could also have led to the collapse of political leadership, as it did in Buffalo. A loss of political leadership, no doubt, would have exaggerated the economic decline. Imagine Pittsburgh as an "accidental" rather than an "intentional" city, and a familiar scenario seems likely: The central city empties out, businesses move to the suburbs or, equally as likely in a region in decline, to other metropolitan areas; the city then is unable to adequately maintain its public infrastructure and services. The Pittsburgh story is remarkable precisely because it does not follow that course. Quite the contrary; the 1985 edition of *Places Rated Almanac* ranked Pittsburgh the most "livable" city in the U.S. (Boyer & Savageau, 1985, pp. 417, 435), mainly because of its high-quality public infrastructure and services and the availability of higher education and medical facilities matched by few other cities. It appears that Pittsburgh was able to substantially direct—if not control—its own destiny.

Elites in Rennes have been as active in promoting economic revitalization as their counterparts in Pittsburgh, but they have been able to use far-reaching public powers that are not available in any U.S. city. Rennes has also benefited from several decades of favorable policies administered by the central state. In contrast to U.S. cities, as Le Galès points out, "private sector elites have not played a major role in Rennes's development." The main actors have been banks and administrative agencies representing the state, and local intellectuals who lead the Christian Democratic Party. Despite the weakness of the private sector, these local elites have long used entrepreneurial skills to promote Rennes's development. They have been successful in securing development money, special projects, and subsidies from the state to support a Science Park, a job-training program, and infrastructure for electronics, communication, health and environment, and bio-industries companies.

The reforms adopted in France after 1983, which have granted local governments unprecedented powers and responsibilities, did not weaken local leadership, which had been accustomed to trying to exact favorable policies from the central government. Quite the contrary. Rennes's elites quickly redirected their entrepreneurial skills to building a constituency to support comprehensive plans to regenerate the local economy. In only a decade Rennes transformed its development priorities to stress the importance of private investment. By the end of the 1980s Rennes's elites had organized a broad-based process of planning to support economic development. To support local firms and jobs, financial assistance to firms is made available, land and advice are offered. The historic center of the city has been renovated, and a regional cultural festival has been organized as part of a marketing campaign to sell the city as a location for conferences and tourism.

The history of economic development efforts in Hamburg shows that the city has moved through three distinct phases. In each period, public policies have been used as instruments to stimulate private investment. The policies available have been far-reaching and interventionist, since Hamburg has had the ability to supply land, direct subsidies to favored industries and sectors, and carry out ambitious urban renewal schemes. All of these policies were sustained through a politics in which government and party elites (the Social Democrats) mobilized broad coalitions embracing labor and trade associations on the left, to business and its allies on the right.

The leadership in Pittsburgh, Rennes, and Hamburg were able to build an institutional and political capacity to sustain economic regeneration because the elites that defined the policies were well entrenched. The political coalitions were not fragile, as they were in Buffalo, Liverpool, and

Vancouver. In Hamburg, industrialization policies that were pursued from 1965 to 1983 proved to be misguided, in the sense that subsidies were poured into industries that continued to decline. But the ability to change course after 1983, to define an "Enterprise Hamburg" program to gentrify the inner city and to define a "from ship to chip" strategy of support for high-tech and tourism and service industries, can be traced to the strength, durability, and adaptability of the coalition supporting regeneration.

Glasgow, Sheffield, Montreal, and Vancouver produced substantial public leadership to support regeneration. But in these cities, public-private cooperation has been somewhat episodic or difficult to sustain. Economic regeneration policies in these cities have emerged recently, and the coalitions that support the policies are still evolving. There is a significant possibility that effective opposition could arise that might divide elites.

Glasgow is the British city that experienced the most profound policy transformation and made the greatest symbolic achievements during the 1980s. As Robin Boyle points out, it is a paradox—a Labour-controlled city that, without changing political control, embraced the strategy of urban regeneration favored by a Conservative government. Within a decade, with a carefully orchestrated and systematic publicity campaign, it projected Glasgow as a city that has thrown off the legacy of decaying traditional industries. It transformed the physical environment in its city center, defined itself a new niche as the cultural capital of Europe, and successfully promoted the growth of a service-sector economy. Leadership in Glasgow has been drawn almost entirely from the public sector, either from the city council or the powerful Scottish Development Agency. More recently, it has engaged the private sector in a public-private partnership with a clear vision of the economic future of the city. And it has developed the capacity to market that vision, first to its citizens inside the city and then to potential public- and private-sector investors elsewhere. All this has made Glasgow into a much-discussed model of urban entrepreneurialism and place marketing.

The resources available to the city's leaders were important. In the first case, it is a long-time Labour-controlled city, which has the advantage of political stability. Glasgow's second advantage was its ability to exploit national public resources. In particular, it could use the Scottish Office, in effect a branch of the U.K. government, to lobby for the city at the central government level and to deliver the financial and administrative resources to underwrite the regeneration strategy in the 1980s. Glasgow's leaders could also exploit a particular form of regional advantage, which Boyle calls the "Scottish community," the tight network of political, professional, and commercial institutions that produced a distinctive, shared view of Glasgow's problems and potential solutions and provided the institutional glue necessary to support a sustained process of change. Unlike Liverpool, Glasgow also has had a relatively substantial private sector with a number of national company headquarters to give weight to private-sector initiative and leadership.

Glasgow is becoming an intentional city. It is one the best examples of the significance of leadership, in that it has been able to maximally exploit its economic, institutional, and social resources. Through the coherence, consistency, and clarity of its regeneration strategy, Glasgow's leadership transformed the image of the city. Yet the last point raises a qualification about its record and an important question for the future. The Glasgow achievement has been possible because of the institutional framework and resources of the public sector at the national, regional, and local levels. Current central government plans to dismantle that well-integrated machinery could throw the regeneration priorities of the Labour administration into doubt and undermine the capacity of the city's leaders.

Sheffield is, in many respects, a miniaturized version of Glasgow, but it started on the same path five years later. It had the local political stability to allow it to switch to new regeneration strategies when it became clear that the municipal socialism of the early 1980s, partly because of the central government's antagonism, was not working. It had skilled local leadership that began to form a public-private partnership and encouraged the private sector to support the achievements of the publicly financed regeneration strategy. It also recognized the importance of place marketing to attract potential investment. It has had a relatively small but stable private sector, which seems dedicated to staying in the city, and it has had the social solidarity of a compact city, which has consistently pursued Labourist welfare policies and avoided substantial internal social and political dissent.

However, the city's position is far from secure. The coalition between the public and private sectors is new and tentative. There are some internal divisions, and it is questionable whether the relationship is well institutionalized. Most crucial, it is not obvious whether, or what, the regeneration strategy will deliver. It embraces fewer economic sectors than Glasgow's. It lacks the economic and cultural resources of Rennes, Pittsburgh, and Hamburg. The strategy remains to be tested by significant political controversy. It is too early to tell if the flagship projects will succeed, or if their benefits will spill over to fuel a generalized revitalization of the local economy. Even if the projects are ultimately successful on their own terms, reservations about their general impact may yet provoke opposition to new projects.

Though Montreal has a relatively long history of redevelopment that dates from the 1950s, the urban elites that have emerged during the past decade have sponsored, in Jacques Léveillée and Robert Whelan's words, "a minor revolution" in economic redevelopment policy. In 1978 a coalition was formed to support a planned regeneration strategy, when, for the first time, business leaders became active participants in defining the city's agenda, and public leaders in turn developed the institutional capacity to coordinate plans and projects. Though Montreal had previously promoted and built infrastructure to support "grand projects" such as the 1967 World's Fair and the 1976 Olympic Games, these activities were essentially opportunistic and were not backed by either development strategies or a stable political coalition. Instead, they were outgrowths of a machine-style politics that proceeded on a case-by-case basis.

The strategies of the 1980s have been, in contrast, explicitly entrepreneurial. They all revolve around promoting Montreal as an international city, a center for international corporate headquarters, finance and trade, high technology and design, and culture and tourism. The political coalition needed to back these strategies is fragile, not only because it is new, but because of potential divisions between English-speaking and French-speaking elites.

The political alliances that have defined Vancouver's regeneration policies have attempted to link concerns about local economic vitality with popular worries that growth might compromise the quality of life that is a unique feature of the city. Until 1986 Vancouver's leftist politicians operated against a backdrop of a conservative provincial politics—not unlike the experience of the Labour-controlled cities under Ms. Thatcher. But Vancouver's "leftist" orientation was never anti-business but corporatist in style. Cooperative relations among business, government, and labor were stressed, and thus the potential hostility between Vancouver and the provincial government remained muted.

Since 1986 Vancouver has been captured by a conservative regime that, according to Warren Magnusson, "put the municipality at the service of land development, and inhibits public participation in strategic planning by capitalist enterprises." In other words, the municipal role in balancing quality of life and development concerns has, for a time, disappeared. But one suspects that this phase may prove temporary. Vancouver's penchant for experimentation with broad-based political movements at the local level has sprung from a local political culture that may well reassert itself. For some time environmental issues have been as important as economic growth in defining the terrain of local politics. It is unlikely that the urban elites who now govern the city can permanently ignore the political necessity for striking a balance between growth and concerns about the quality of life.

But the capacity for such a role will have to be rebuilt. The provincial government has imposed policies of fiscal austerity that have limited the ability of local authorities to deal with social and economic problems. Funding for social services and housing programs has been reduced. In 1983 the province stripped regional districts of their planning powers to make it easier for developers to escape regulation. The election of a conservative mayor and council in 1986 has led to the dismantling of the city's attempts to regulate development and promote citizen participation in planning and economic development. The "strategic initiatives" approved by Vancouver's council in 1983 to make the city a relatively autonomous center of global capital have been abandoned. The guided development of the 1970s and early 1980s has been replaced by a regeneration process that relies on fiscal restraint, weaker unions, and a business environment free from external controls.

Baltimore's version of economic regeneration has combined a rich and complex institutional structure with a weak and passive public leadership and autonomous private-sector control. As Richard Hula shows, though the city's long-serving mayor, William Schaefer, was a tireless advocate of redevelopment, he was also a fiscal conservative, skeptical that public bureaucrats could play a useful role in regenerating the city. The redevelopment initiatives that changed the face of the city between 1971 and 1986 relied almost entirely on the private sector for initiation and implementation. To facilitate this mode of policy-making, a large number of quasi-private corporations were established, most of them devoted to carrying through a single project. Each of these organizations has received significant public funding and subsidies; they exercise significant public powers such as eminent domain; and they exert a great amount of influence on capital spending in central Baltimore. Yet they operate as private corporations, jealously guarding financial records, procedures, and policies as private property. Their far-reaching powers and activities, combined with their mode of operation, make them a "shadow government" that, in many ways, makes the most important decisions affecting Baltimore's citizens.

The creation of a shadow government flowed as a matter of course from a distrust for municipal bureaucracies and a conviction that private-sector institutions would only invest their money in Baltimore if they controlled the renewal process. The marginalization of elected officials and public bureaucrats also indicated a disdain for democratic politics, whose fractious nature might interfere with efficient program implementation. Negotiations among developers and other participants has proceeded entirely behind the scenes.

Within the past two years, political pressures have built to open up the process. It is clear that redevelopment has created "two Baltimores," one that revolves around and benefits from redevelopment—financial institutions, corporations, white-collar professionals—and the other Baltimore of dilapidated neighborhoods that receive underfunded city services. A rhetorical move has been made by the city to assert public control over physical redevelopment. It remains to be seen whether there actually will be meaningful policy shifts.

From Houston's founding to the present, business leaders have dominated the politics of the city, treating local government as their private preserve. Houston has rarely experienced significant pressures to create a more inclusive style of politics. Local business elites historically have been able to exercise autonomous control over nearly all governmental decisions, and recent efforts at economic regeneration reflects this history. The Houston Economic Development Council (HEDC), a 1984 creation of the Houston Chamber of Commerce, has presided over all of Houston's economic development policies. The HEDC is organized along the same principles as the quasi-private organizations that make up Baltimore's shadow government. Though about one-third of its funding comes from contracts with local public agencies, it is run as a private corporation. The HEDC executives jealously guard their private status and have often expressed worry that HEDC records might become public.

Although Houston offers an array of tax abatements and direct subsidies, the main strategy for economic regeneration has been advertising Houston's "good business climate." To local business elites, the principal element of this climate is a weak local state. Almost by definition, this has meant that policies of regeneration are passive and rather broad, rather than active, targeted, and carefully administered. In the political climate of Houston, the only way to build a capacity to support ambitious and targeted policy would be to follow Baltimore's model of a shadow government of quasi-private agencies. Indeed, this would be a logical next step for business elites in Houston, for it would allow them to build an institutional structure to support ambitious and targeted regeneration strategies without significantly expanding the local state.

Liverpool, Marseilles, and Buffalo are the cities that have the most obvious leadership deficits. The recent histories of Liverpool and Marseilles have striking parallels—economic decline, social division, and political confusion. Liverpool has suffered massive structural decline over the past two decades, but it has generated only limited growth in high-value sectors of the local economy. In the 1980s it experienced social discord and serious rioting. In the early years of the decade its leadership made disastrous choices, and since then the city has not been able to build a coalition to back coherent regeneration strategies. Its public and private sectors are relatively weak. The city government is poorly managed and inefficient, and held in low esteem by the private sector. The private sector itself suffers from the branch-plant syndrome and has produced few powerful leaders who are willing or able to commit their personal or corporate resources to civic life.

Liverpool has failed to maximize it potential resources. It has had poor relations with the central government for more than a decade, and thus it has failed to either extract or exploit the kind of national support that laid the foundation for Glasgow's economic revival. And Liverpool is a difficult community to mobilize. In sharp contrast to Rennes and Pittsburgh, for example, economic and demographic decline have left it with a predominantly working-class culture, powerful left-wing and trade unions, considerable industrial militancy, and few economic and cultural migrants. Typical educational levels are low. Even a powerful leadership would find it difficult to mobilize the various segments of the community behind a well-defined strategy of economic renewal. A fragile public-private sector leadership is attempting to generate support for an economic strategy that is broader than a public infrastructure program. However, the results of the minor economic renaissance that has taken place in recent years have been unevenly spread, and there remains local skepticism about whether the strategy will meet the needs of the majority population. When this is added to the city's recent record of economic and political instability, there can be no doubt about the scope of the challenge facing the city's leadership in 1990s.

The port-based economy of Marseilles went through a boom-and-bust experience not unlike Liverpool's. Municipal government has historically been weak, inefficient, and notoriously corrupt. Reflecting its immigrant character, the city has always been highly segregated. In the past decade, an insurgent proto-fascist, anti-immigrant party has mobilized substantial support, and as a result, local politics have been highly contentious. The city remains extremely divided in its social structure.

In contrast to Liverpool, however, a powerful leadership, which had excellent connections to the central government, ruled Marseilles for many years. Behind the machine-style leadership of Gaston Defferre, the city was able to modernize its infrastructure during the 1960s and 1970s. But the period of most severe economic crisis coincided with a political crisis. The gradual disintegration of Deffere's political machine, climaxed by his death in 1986, introduced considerable political instability to the city. But a new leadership is being created in Marseilles that may mobilize not only political constituencies but economic leaders as well.

Buffalo is a case of a city that experienced the collapse of not only its local economy but its leadership as well. The absentee ownership of local and regional enterprises resulted in a leadership vacuum in the private sector. Political leaders tended to focus their attention on the spoils of a declining city. In late 1983, when the new publisher of the *Buffalo Evening News* initiated a series of dinners to build a civic leadership, he found himself in the position of attempting to identify the city's key leaders. Ultimately, when the "Buffalo 18" finally was convened, the weakness of local public institutions was so obvious that the group devised, in effect, a strategy of importing bureaucrats and professionals from other cities. This was a logical step, especially considering that one-half of the members of the Buffalo 18 themselves came from outside the region. This group has intervened in the selection of almost all public and not-for-profit administrators in an attempt to build a cadre of professionals and an institutional fabric with the capacity to provide leadership for the city and region. As David Perry points out, the Buffalo 18 is trying to "recast" a leadership, almost like selecting actors for a play. In this process, elected officials have become less and less important. Local democracy has become devalued in part because elected leaders are, at present, simply not considered as relevant actors.

THE CONSEQUENCES OF REGENERATION

Global capitalism forces cities to compete for investment, but it does not "determine the form in which . . . resources are invested" (Fainstein, 1990, p. 25). Nevertheless, the options for cities in the last decade of the twentieth century seem to be limited to permutations on a service-sector economy, as Susan Fainstein points out in her essay. This point is intimately related to how the benefits of regeneration get distributed. Service-sector growth creates a severely segmented labor force, and this is as true in West Germany, France, Britain, and Canada as it is in the United States. People with years of formal education or special training take the desirable jobs that are created when regeneration is successful. The multitude of service-sector jobs at the other end tend to pay minimum wage, often are temporary, and carry few benefits. And in any case, these jobs have not actually been created fast enough to replace the jobs lost through deindustrialization. It is remarkable that unemployment rates in Montreal hover close to 20%—about the same level as in Liverpool and Glasgow.

The existence of a strong leadership in support of regeneration, even if it relies on the mobilization of diverse political groups, does not guarantee that the benefits of regeneration will be equitably distributed. In Pittsburgh, the people thrown out of work by deindustrialization are not the same people who qualify for the well-paying jobs in the hospitals, medical laboratories, research parks, or universities. Indeed, the victims of deindustrialization will likely find themselves displaced by the gentrification of their neighborhoods. In Rennes, urban elites have attempted to build an inclusive politics that incorporates most of the groups that would be affected by economic change. Nevertheless, the structure of the new economy favors people with high education and job skills. Unemployment remains a problem for those without those skills and is unlikely to be solved by the new growth sectors. In Hamburg, the poor have been politically marginalized and have borne the brunt of the new policies, as Jens Dangschat and Jürgen Ossenbrügge have demonstrated.

The case studies in this volume lead us to conclude that leadership may make a considerable difference in whether, and how, a city regenerates its economy—but may make much less difference in addressing issues of equity. The task of equitably distributing, as well as creating, wealth has eluded most cities, which moves us to ask whether current definitions of economic regeneration are sufficient. Over the past few years the term has, in the main, been appropriated by conservatives and given a narrow meaning. It is not difficult to imagine a more expansive definition of regeneration, which would concern itself with wealth distribution as well as wealth creation, with investment in human capital as well as physical capital, and with the losers as well as the winners in the regeneration process. But such a definition would become convincing only if the political calculus changes. As Susan Fainstein points out in her essay, those who favor the equitable distribution of the benefits from growth must also have a vision that simultaneously embraces growth and economic change. If such a vision were to energize new political movements and coalitions in national politics as well as in cities, then, and only then, would there exist a greater range of choices for urban leaders seeking to regenerate their local economies. Until a broader definition of regeneration is realistically available, the limited versions described in this book will constitute the options available to local elites. However, even given these limited choices, success remains better than failure. Leadership creates the possibility for success, defined as more investment in a community. Perhaps in the future it also can become the means of changing our conception of success.

REFERENCES

Boyer & Savageau. (1985). *Places rated almanac*. New York: Prentice-Hall Press.

Fainstein, S. (1990). *Urban economic development and the transformation of planning in the United States and Great Britain.*

About the Contributors

ROBIN BOYLE is currently Senior Lecturer and Head of Department at the Centre for Planning, University of Strathclyde, Glasgow, UK. Comparative urban policy is one of his central research and teaching interests and he has published widely on this topic. His most recent book, *Privatism and Urban Policy,* was coauthored with Tim Barnekov and Dan Rich (Oxford, 1989). This book was the product of a Visiting Professorship at the University of Delaware in 1985 and underscores the close collaboration between colleagues from the Universities of Strathclyde and Delaware.

JENS S. DANGSCHAT is Assistant Professor in the Department of Sociology, University of Hamburg (FRG) and Vice-Director of the Centre of Comparative Urban Research. He has published numerous articles on comparative urban development in Western and Eastern Europe, residential segregation, new household types, gentrification and urban development of Hamburg. He is the author of *Soziale and räumliche Ungleicheiten in Warschau* (Social and Spatial Disparities in Warsaw) and is coediting three books on gentrification (in Hamburg, in the FRG, and in Europe).

ANDRÉ DONZEL has lived in Provence for 20 years. He has been a sociologist at the CNRS since 1980 and works in the Centre de Recherche en Ecologie Sociale, a research unit of the Ecole des Hautes Etudes en Sciences Sociales in Marseilles. His research mainly focuses on urban life and politics in Marseilles.

SUSAN S. FAINSTEIN is Professor of Urban Planning and Policy Development at Rutgers University. Among her books are *Urban Political Movements, Urban Policy under Capitalism,* and *Restructuring the City.* She is currently working on a book comparing planning and urban development in London and New York.

JOE R. FEAGIN is Professor of Sociology at the University of Florida. Author of two dozen books and several dozen scholarly articles, he does research on public policy and inequality issues, particularly race and gender discrimination, and on urban and community development, including comparative studies of European and American cities and urban planning. A product of this latter research on development is *Free Enterprise City: Houston in Political-Economic Perspective* (Rutgers University Press, 1988). Some of his theorizing about comparative urban research has appeared in *The Capitalist City* (Basil Blackwell, 1987), edited with Michael Peter Smith. He has recently completed work on economic and political development and economic decline in Houston and Aberdeen, the "oil capitals," with a particular focus on planning responses. Another current research study, funded by the Will Hogg Foundation, involves the difficulties and discrimination faced by successful Black Americans in U.S. cities.

RICHARD C. HULA is an Associate Professor in the Institute for Urban Studies at the University of Maryland at College Park. His current research interests center on the use of the private sector to implement public policy. He has published a number of articles on various aspects of privatization, including economic development and public housing.

DENNIS JUDD is Professor and Chair in the Department of Political Science, University of Missouri-St. Louis. He has published extensively on urban political economy, urban revitalization, and public policy. Recent books include *The Politics of American Cities: Private Power and Public Policy*; *The Development of American Public Policy* (with David Brian Robertson); and *Regenerating the Cities: The U.K. Crisis and the U.S. Experience* (coedited with Michael Parkinson and Bernard Foley). He is coeditor of the *Urban Affairs Quarterly.* He continues to conduct comparative urban research.

PAUL LAWLESS is a Reader in Urban Policy at Sheffield City Polytechnic, where he has taught since 1977. Prior to that date he lectured at the Architectural Association. He has written numerous books and articles on local economic development and British inner-city problems. His most recent book, *Britain's Inner Cities,* was published in 1989 (Paul Chapman Publishing). In 1988-1989 he was awarded a Nuffield Foundation Fellowship to examine aspects of managed workspaces in the United Kingdom.

PATRICK LE GALÈS is a Research Fellow at the Observatoire Sociologique du Changement CNRS-Fondation Nationale des Sciences Politiques, in Paris. He was educated at the University of Paris, Nanterre, and at Nuffield College, Oxford, where he completed an M.L.H. on local economic policies in France and Britain. He is currently working on issues of comparative urban economic development and has written about a wide variety of cities in France and Britain, as well as about national urban policy in those two countries.

JACQUES LÉVEILLÉE teaches political science at the University of Quebec in Montreal and specializes in urban politics. He is coauthor of *Montreal After Drapeau* (Black Rose Books, 1986) and *Le Systéme Politique de Montréal* (ACFAS, 1986). He is currently studying local political parties in Toronto, Montreal, and Quebec City on a comparative basis.

WARREN MAGNUSSON is Associate Professor of Political Science and director of the Interdisciplinary M.A. Program in Contemporary Social and Political Thought at the University of Victoria, British Columbia. He is a graduate of the University of Manitoba and the University of Oxford. His books include *City Politics in Canada* (with Andrew Sancton) and a forthcoming study of "radical municipalities" in Britain and North America.

JÜRGEN OSSENBRÜGGE is Assistant Professor in the Department of Geography, University of Hamburg (FRG). He has published extensively on problems of regional restructuring and spatial policies, locational conflicts, and geopolitics. He is now completing a book on "Environmental Risk Perception, Social Movements and Regional Change" and is coeditor of *Polarized Development and Decentralization Policies in Central America.*

ROBERT E. PARKER is Assistant Professor of Sociology at the University of Nevada, Las Vegas. He is the author of several papers on urban-related issues and is coauthor with Joe R. Feagin of *Building American Cities* (1990). He recently received a Nevada Humanities Committee Grant to study the social costs of rapid urban growth in Southern Nevada.

MICHAEL PARKINSON is Director of the Centre for Urban Studies, the University of Liverpool. He has written extensively about educational and urban policy in Britain. His recent books include *Liverpool on the Brink* and two edited volumes, *Reshaping Local Government* and *Regenerating the Cities: The U.K. Crisis and the U.S. Experience*. He and staff members of the Centre continue to conduct comparative research on urban regeneration, funded by the Leverhulme Trust.

DAVID C. PERRY is the Albert A. Levin Professor of Urban Studies and Public Service at the College of Urban Affairs, Cleveland State University. He has published extensively on urban economic development and the politics of planning. He is presently involved in a multiyear study of Robert Moses and the use of public authorities in urban governance. One of his books is *The Rise of the Sunbelt Cities* (edited with Alfred Watkins).

ALBERTA M. SBRAGIA is an Associate Professor of Political Science and the Director of the West European Studies Program at the University of Pittsburgh. She has edited the *Municipal Money Chase: The Politics of Local Government Finance* (Westview, 1983) and authored articles on various aspects of finance and public policy in Italy, France, Great Britain, and the United States. She is currently writing a book on the relationship between law and capital markets in the area of capital investment. She is also directing a project for the Brookings Institution on the institutional implications for the European Community of the "1992" program.

ROBERT K. WHELAN is Professor and Associate Dean in the College of Urban and Public Affairs at the University of New Orleans. He is coauthor of *Urban Policy and Politics in a Bureaucratic Age* and author of numerous articles and papers on urban economic development. His major research interest is urban economic development, with a recent focus on the New Orleans and Montreal metropolitan areas.